THE STEPS OF CREATION
PART II: COSMOS

Clare V. Merry

Published by The MerryScience Foundation

Illustrations by Clare Merry

ISBN: 978-1-909204-82-9

The MerryScience Foundation
is an imprint of
Dolman Scott Ltd
www.dolmanscott.com

TABLE OF CONTENTS FOR PART II

Chapter 13 STAR FORMATION AND EVOLUTION

Chapter 14 THE STRUCTURE OF MATTER

Chapter 15 THE EXPANDING UNIVERSE

Chapter 16 GALAXIES: EDDIES IN THE UNIVERSE

CONCLUSION II

APPENDICES

LIST OF ILLUSTRATIONS FOR PART II

The computer graphics, ink and pencil drawings in this book are by the author. A few have been adapted from other sources where indicated. Tables have been compiled by the author using data from various sources.

COVER DESIGNS

The pastel drawings by the author for the covers of Parts I, II and III of *The Steps of Creation* represent the early stages of planet Earth when it was inhabited only by microbial life.

The cover of Part I shows the Earth inoculated with chemolithotrophic methanogen Archaea. The Earth is in the darkness of outer space when these microbes start to produce water through biochemistry upon the dry basalt rock surface.

On the cover of Part II Purple bacteria bloom in pools of water on Earth's cratered surface. These anaerobes capture rays from the dim juvenile Sun practicing anoxygenic photosynthesis, and thereby increase the amount of water on Earth. The surface is strewn with the blackened products of anaerobic metabolism and elemental sulphur.

The cover of Part III depicts the dawning of Earth's aerobic atmosphere as the Earth is colonized by cyanobacteria whose photosynthesis releases free oxygen. The shallow seas are lined with bacterial mats and the bare rock surface is colonized – everywhere is green, tinged with the red of rusted iron.

Scenarios of the early stages of planet Earth are given in the following chapters of *The Steps of Creation*:

The stages of planet Earth are described in Chapter 10, section 10, of Part I.

Candidates for life in space are given in Chapter 11, section 7, of Part II.

The stages of proto-planet Earth are described in Chapter 18, section 4, of Part III.

THE THEORY OF NANOCREATION AND ENTROPIC EVOLUTION

Manifesto

The central idea of this theory is that God created at the microscopic, nanoscale level, and then allowed macrostructures to evolve naturally. Entities at the nanoscale level include such things as the hydrogen atom, certain elements, organic molecules and DNA. Macrostructures include such things as galaxies, stars, the Earth with its biofriendly environment and our own bodies. The term 'nanocreation' is adapted from the new science of nanotechnology.

Nanotechnology involves the control of the structure of matter on a scale below 100 nanometres. One nanometre is 10^{-9} metres or a millionth of a millimetre. Nanotechnology is performed using ultraviolet and X-ray optical techniques or an electron beam to position groups of atoms or molecules in an engineered material. This human creative activity may serve as an analogy for the creative activity of God.

Genetic engineers now make transgenic mice by inserting genes from humans into the mouse genome for the purposes of research. This feat of molecular manipulation is performed with pre-existing genes. It is proposed that God can equal human creativity, and better; He could make a new gene that would encode a new protein as a component of life.

Entropic Evolution is inspired from the principle behind the Second Law of Thermodynamics. An increase in entropy is a move away from order towards a system characterized by disorder and inertia. The Laws of Thermodynamics relate to energy, whereas the entropy involved here relates to information; this is known as Shannon Entropy since it was first proposed by Claude Shannon in 1948. The term Entropic Evolution is used to convey the idea that, although evolution gives rise to interesting phenomena, it is in itself a winding-down process.

The Theory of Nanocreation and Entropic Evolution has some hypotheses which seek to define primary causes – the intervention of God in the universe at specific points in time and space, while other hypotheses seek to explain secondary causes – the working of natural evolutionary processes, and their limits. The combination of these primary and secondary causes gives rise to a global scheme. It is unashamedly theistic science.

Nanocreation does not involve a literal reading of the first chapters of Genesis. The theory accords with the essence of the creation account of Genesis, and its underlying truths.
The essence of the Genesis story is that God is the Creator of both matter and life. The days of creation convey the idea that creation occurred as a sequence of acts of creation over time, culminating in the creation of humankind.

The Theory of Nanocreation and Entropic Evolution may be accepted as a scientific theory for the following reasons:

- It is based on data generated by modern, orthodox science and thus shares common ground with other current scientific theories.
- It is a theory composed of hypotheses, not just a literal interpretation of the Bible.
- It has generated predictions that are testable by science, and which lay it open to being proved wrong in part or as a whole; it is a falsifiable theory.

The theory incorporates an explosive beginning to the universe which I have named the Multi-Bang. The Multi-Bang has some important differences with the Big Bang Theory. Energy in the form of motion in created matter gave rise to an expanding universe.

In line with main stream science, the universe is accepted to be around 13.7 thousand million years old. The timescale for the appearance of the first life forms on Earth is counted in thousands of millions of years. It is accepted that radiometric dating places some of the earliest microbial fossils – stromatolites from the Precambrian era at 3.5 thousand million years old.

Evolution is viewed as something positive; its possibility incorporated in the original set-up. Evolution is natural in that it consists of unguided processes happening by chance. There is both cosmic evolution involving the 'heavenly bodies' and biological evolution occurring in the different forms of life.

Charles Darwin presented his Theory of Descent with Modification through Natural Selection in *On the Origin of Species* in 1859. The concept of the modification of pre-existing traits through Natural Selection giving rise to diversity in the organic world is fully incorporated into the Theory of Nanocreation and Entropic Evolution. Darwin mainly establishes the Theory of Common Descent to class level; this is also endorsed by the Theory of Nanocreation and Entropic Evolution. Another meaning of the term Natural Selection expanded by Darwin in the second half of *The Origin of Species* is the step-by-step building-up of new structures and organs representing an increase in complexity, and couched in terms of usefulness. This strand of evolutionary thought is rejected by the Theory of Nanocreation and Entropic Evolution.

The concept of Entropic Evolution in the biological world rests on the well-documented observation that mutation causes dysfunction in the genetic system. Mutation tends to switch genes off such that they are not expressed. However, what is dysfunction in the genome often manifests itself in the organism as modification and variation. Where variation proves useful, it is preserved by Natural Selection. Many examples of evolution, including the classic examples, in fact, involve loss of structures and traits in the organism. Thus, the selection of variations does not build up complexity as Darwin claimed, but it does produce adaptation to the environment and a great diversity of species.

NeoDarwinism emerged in the 20th century by combining Darwinist theory with the laws of inheritance, and the discovery of genes and DNA. A new idea was added – that of the origin of life from non-life by natural processes involving selection. A new philosophy came to dominate science asserting the incompatibility of science and religion.

Mainstream science today has an ethos of rationalism and objectivity. However, many materialistic hypotheses offered within the context of science to answer the big questions are a matter of belief, not of scientific demonstration. They are deemed worthy of science textbooks purely on the basis that they do not involve any interaction between a Creator and creation; it is thought that their materialism makes them 'scientific'. Science today follows the Philosophy of Naturalism – a philosophy unknown to the founders of modern science. Naturalism accepts only natural explanation for the working *and* origin of all phenomena, and excludes God from its definition of reality. This *a priori* stance does not make the many scientific theories necessarily objective, nor even rational.

The Theory of Nanocreation and Entropic Evolution is a rejection of purely materialistic science and the Philosophy of Naturalism. It is science which does not exclude God. To the debate that rages between Creationists and Evolutionists, the Theory of Nanocreation and Entropic Evolution offers a middle way: creation in the context of an evolving world.

Introduction

Parts I and II of *The Steps of Creation* are about the Theory of Nanocreation and Entropic Evolution. Part III contains hypotheses relating to natural processes of wide application and especially Earth history.

In *The Steps of Creation* I will present the current theories and hypotheses of science – that is, the interpretation of the facts which are given in science textbooks, popular science programmes and encyclopedias. I will present my alternative theory and hypotheses as a different interpretation of the same facts. I will also discuss some ideas, notions and views that are either religious or philosophical, and which relate to science and the natural world.

The Theory of Nanocreation and Entropic Evolution is composed of nine hypotheses; six hypotheses are presented in Part I and three in Part II. There are three additional hypotheses in Part III which are not part of the Theory. *The Steps of Creation* is presented in three volumes corresponding to the three parts: Part I Life; Part II Cosmos; and Part III Earth.

The hypotheses in Part I relate to Nanocreation and Entropic Evolution. These hypotheses concern the DNA code and genetic system, the classification of unicellular life, plants and animals, the origin of life, the formation of ecosystems, and the origin of water and the conditions for life.

The hypotheses in Part II relate to Nanocreation and cosmic evolution. The hypotheses of these chapters relate to the origin of the elements and extraterrestrial life, the formation of stars, subatomic particles, the explosive beginning to the universe, and the formation of galaxies.

The hypotheses in Part III concern Earth history. The hypotheses of this last part propose mechanisms that are natural processes bringing about the formation of the solar system and Earth as a planet, and moulding the features on the surface of the Earth. In Part III Earth history is presented in a way it has never been presented before.

Some readers may accept the whole Theory of Nanocreation and Entropic Evolution, while others may accept some hypotheses and not others. It is my view that we should be free to choose what we believe and free to question so-called 'proven facts'. Science has never advanced without questioning. The Theory is offered in the spirit of debate.

Part I of *The Steps of Creation* covers the following subjects:

- Genetics (also called molecular biology)
- Biology and some human biology
- The classification of organisms (also known as taxonomy or systematics)
- Palaeontology
- Origin of life theories
- Biochemistry
- Microbiology

Part II of *The Steps of Creation* covers the following subjects:

- Astronomy of stars, interstellar clouds and planets
- Astrophysics (the physical and chemical constitution of stars)
- Chemistry
- Physics
- Particle physics
- Cosmology (the origin and development of the universe)

Part III of *The Steps of Creation* covers the following subjects:

- Astronomy (of the solar system)
- Geology
- Earth sciences

A basic understanding of each of these subjects will be given in terms of facts and current theories. At the same time, the hypotheses I present belonging to the Theory of Nanocreation and Entropic Evolution, will show how the facts can be understood in new ways and in relation to the 'Steps of Creation'.

This book will not include details concerning the scientific methods used in investigation or the way in which facts are derived from raw data. There will be no explanations of the workings of scientific instruments. There are many good introductions to science which explain the methods used very well. At most, I will mention the name of the method used in the relevant scientific investigation, only in passing. The reader can use this to look up the methods in an encyclopedia if they wish.

The aim of this book is to get an overview of the facts generally accepted as such by science, and with knowledge of these facts seek further understanding.

One could engage in a long discussion on the theology and philosophy arising from this book, but I do not think it would be profitable. I do not wish to wander too far from the physical details of the natural world.

The English spelling used in this book is from *English Dictionary* Geddes & Grosset (1999) New Lanark, Scotland. The choice of 'z' rather than 's' accords with older English spelling. The use of 'z' is more acceptable to the Americans, and I prefer it.

I will use 'thousands of millions', rather than 'billions'. This is because many people do not know what a billion is. In Britain a billion has been, until recently, a million million, while in the USA it is only a thousand million. The words 'thousand million' make it clear what we are talking about and avoid confusions with zeros.

This book will not be over-burdened with mathematical equations. As scientific words are introduced, they will be explained in the text. Scientific jargon is avoided where possible.

In *The Steps of Creation*, hypotheses, theories and models in science will be given capital letters. My own theory and the hypotheses of which it is composed will also be given capital letters. Likewise, philosophies and some concepts will be capitalized. This is because this book is about ideas and concepts in science. Ideas, notions, concepts and ways of viewing things are to be distinguished from known, proved and agreed-upon facts. A fact will be taken as a fact when there is no longer any dispute as to its existence or veracity.

Normally, Natural Selection is not given capital letters in science books, and although a concept, it is treated as a fact. I have used capitals for Natural Selection to emphasize that it is a concept with various possible meanings. The various possible meanings of Natural Selection are central to the discussion contained in *The Steps of Creation*.

If I have capitalized Natural Selection, should I have capitalized evolution? Is evolution fact or theory?

The word evolution has a much wider application than the term Natural Selection. Sometimes we are all agreed that something has evolved, and so it presents itself as a fact. In other instances, the term evolution competes with other explanations whether religious or philosophical. In this case, evolution presents itself as a belief or theory used to explain nature, and it would be clearer to write it as 'Evolution'. I finally decided to keep evolution without capitals due to the ambiguities involved. I, likewise, decided to leave creation without capitals, although all sorts of discussions about what we mean by 'creation' could be had (I make an exception to this rule in Chapter 5 because this chapter is a discussion of ideas).

In our society evolution is part of our world-view and dominant philosophy. The word 'evolution' or 'evolved' is constantly slipped into common speech, often where it does not belong. A phrase often used is "the evolution of life on Earth", when what is meant is "the existence of life on Earth" since we do not finally know where life came from.

Stars evolve, life evolves and humans have evolved. This may be true, but it still does not explain the origin of these things.

The assumption is that things, in fact, everything around us is the way it is because it evolved that way. It is the unexamined assumption that puts an end to further questions. This is the world in which we live and which we must examine – if we can just stand back from it a bit.

The Steps of Creation is a book about agreed facts, and debatable theories and concepts. It offers new scientific interpretations of the facts. It presents a reality behind which is God – Creator, Sustainer and Father of all.

Natural Theology, traditionally seeks to prove the existence of God through the observation of nature. This has been done through expounding the design found in nature, and the sense of balance and harmony in the natural order. I am not trying to prove that God exists. I am sharing my views on creation with fellow believers who believe in God for their own personal reasons, and because of their own experiences of spiritual things, and with anyone else who cares to listen.

God is Being, and so requires no justification for His existence other than Himself. Science cannot ultimately prove or disprove the existence of God, since God is beyond the physical realm. However, because God is a Person, He can prove His own existence, as and when, He chooses.

Thus, the existence of God is taken as given as regards the Theory of Nanocreation and Entropic Evolution.

I seek an understanding of nature as a search for truth and a valid enterprise in its own right, not to prove to anyone that God exists. I believe that God does not wish us to prove His existence with science, because if He had wanted us to do so, He would have made it easier.

My objective is to explain nature, but my scheme only makes sense if there is a Creator God. To Him be the glory.

8

CHAPTER 11

LIFE IN SPACE: INTERSTELLAR CLOUDS

Introduction

In Part I of *The Steps of Creation* we looked at: the genetic code of life; the classification of organisms at different levels of complexity; the origin of life; and aspects of the formation of a biofriendly environment on Earth mediated by various forms of unicellular life. In Chapter 11 – the first chapter of Part II – we are going to turn to look for the location of the first forms of life to come into existence. It is my belief that the first forms of life originated in interstellar clouds and continue to populate interstellar dust clouds today.

The scientific evidence for holding this belief will be explored in detail in this chapter. The hypothesis that I propose stating this belief will be formulated in Chapter 12 along with further evidence.

Chapter 11 is one of the most lengthy of the whole book; however, the details contained in it will give the reader a very good overview of what is currently known about the sorts of clouds contained in the interstellar space of galaxies.

The chapter looks at interstellar clouds, molecular clouds and dark nebulas. The composition of interstellar dust will be discussed. Comparisons will be made with the composition of comets passing through the solar system and meteorites that have fallen to Earth. The rings of dust orbiting planets in the outer solar system will also be included in this chapter.

After all the Earth-bound ideas and theories on the origin of life, it is astounding to find that interstellar space is full of organic molecules, even complex ones. The question we are going to ask is; what on earth (or in space) could they be doing there?

I propose in Chapter 11 that there is an abundance of unicellular life far beyond the confines of the Earth. This is a vision of the vastness of creation.

1. Interstellar clouds

Radioastronomy, developed in the 1950s, provided the means of studying the diffuse matter found between stars. A method known as spectroscopy is used to identify the atoms and molecules that compose interstellar clouds.

The composition of the universe is 99 % hydrogen and helium. The interstellar clouds found within galaxies are composed of 90 % hydrogen atoms, 9 % helium atoms and less than 1 % heavier atoms. A typical interstellar cloud has a total mass several thousand times the mass of the Sun. Hydrogen takes various forms within interstellar clouds.

Diffuse interstellar medium

The diffuse interstellar medium consists of neutral atomic hydrogen. The density of these atoms is very low, only about a hundred atoms per cubic centimetre, and the temperature is between 10 and 100 K with an average of 70 K (−203 °C). There are also atoms of sodium, potassium, ionized calcium and iron in the interstellar medium. In our galaxy, The Milky Way, about half of the hydrogen is atomic and is concentrated in the plane of the galaxy. In between interstellar clouds there is even lower density atomic and ionized hydrogen.

Ionized hydrogen

Ionized hydrogen occurs in regions around hot stars where ultraviolet radiation emitted by the star strips off electrons from hydrogen atoms in a well-defined ionization zone. Ions are charged atoms, as opposed to neutral hydrogen atoms. Circumstellar zones of ionized hydrogen are called H II regions meaning hydrogen with one electron removed. (Neutral hydrogen regions are labelled H I). These regions with intense UV and far-UV radiation causing photoionization of atoms extend for a few tens of light years around stars and absorb most of the photons emitted. Ionized hydrogen has a temperature of 10 000K.

Molecular clouds
A large proportion of hydrogen in galaxies is molecular (H_2). The density of molecular clouds is ten times higher than that of the diffuse interstellar medium with over a thousand atoms per cubic centimetre. Molecular hydrogen is found where the interstellar clouds are dense enough to provide screening from ultraviolet radiation emitted by stars which would dissociate the molecules. Giant molecular clouds form in areas that show a transition from the very cold H I regions of atomic hydrogen and the very hot H II regions of ionized hydrogen. The core of the molecular cloud may have the frigid temperature of 10 K, but in heated interstellar clouds the hydrogen molecules emit infrared lines indicating a temperature of 2000 K.

Dust clouds
Molecules in space are intimately allied with dust; wherever one finds a molecular cloud, one finds lots of dust. Dust grains represent only 1-2 % of the total mass of a molecular cloud, but they are concentrated in some regions. Astronomers give the name dark cloud or dark nebula to an interstellar cloud that contains so much dust that it blots out the light of stars within it and behind it.

Thus, a molecular cloud has two parts: a large lower density cloud surrounding a dense, small core. The molecular cloud may have a few hundred solar masses, while the core may have only five solar masses and so be much smaller in extent. Enormous molecular clouds have been observed of up to 10 000 solar masses. The temperature of the core may be 10 K and the density as high as a million molecules per cubic centimetre (10^{12} molecules per cubic meter).

2. Dark nebulas

We have seen that interstellar clouds are massive conglomerations of mainly atomic, ionized or molecular hydrogen. Interstellar clouds may be overall dilute, but they are highly inhomogeneous in both density and composition. Clouds of molecular hydrogen contain dark cores which are full of dust. The dust is associated with many types of organic molecules which we will examine shortly.

In the rich molecular region of Orion there emerges a dark nebula in the shape of a horse's head which obscures the stars behind it. In localized regions such as the Horsehead Nebula the density reaches one million or more hydrogen molecules per centimetre cubed, compared to an average density of one atom per centimetre cubed for interstellar space overall. Interstellar dust makes its presence known both by the dimming and reddening of starlight. It also makes itself known by the reflection of light in reflection nebulas.

Reddening of light

When the Sun sets on Earth's horizon, it is often tinged with red. The reason for this is that the light from the setting Sun travels through more atmosphere than when the Sun is overhead. Dust in Earth's atmosphere scatters the shorter wavelength blue light leaving the longer wavelength red light. Greater numbers of red light waves therefore reach the viewer, making the setting Sun and sky appear red. Interstellar dust also scatters blue light and so has a reddening effect on starlight coming from behind the cloud to the astronomer viewer on Earth.

Thus, one effect that dust particles have is to scatter light, sending the light waves off in different directions from the original one. Stars viewed through interstellar clouds are reddened and this colour effect can be measured to estimate the amount of dust in a cloud.

Reflection

When interstellar dust scatters light passing through the cloud, the scattered blue light bounces around the molecular cloud until it finally exits, and this can give the cloud itself a bluish appearance. This phenomenon produces a bright cloud with a bluish colour called a reflection nebula. An example of a reflection nebula is the nebula surrounding the Pleiades.

Dimming

Another effect that dust particles have is to absorb light. Absorption of light by interstellar dust causes the dimming of stars located behind or within molecular clouds. The scattering and absorption of light cause less light to exit a molecular cloud than entered it. The dimming of starlight in dark nebulas is known as extinction.

Dark nebulas such as the Horsehead Nebula in Orion display dramatic cut offs of light due to the density of dust they contain.

3. Interstellar dust grains

Interstellar dust is composed of small, solid particles called grains that emit infrared radiation.

The consensus today is that interstellar dust is mainly composed of silicates and water in the form of ice. The cosmic abundance of certain elements suggests that methane, carbon dioxide and ammonia may also be associated with interstellar dust (Zeilik 2002, page 316). Iron may be found in the core of dust grains. Diverse carbonaceous compounds are mixed into the dust grains and different molecular clouds show different mixtures of these organic molecules.

Nobody knows what an interstellar dust grain looks like, but a popular model shows a small core composed of silicate, iron or graphite surrounded by a mantle of ice composed of water, carbon dioxide, methane and ammonia covered with organic tar-like molecules on the outside.

It is known that the amount of ice surrounding the core varies with the temperature of the molecular cloud. In cooler conditions the ice accumulates, while in warmer conditions it may evaporate away, leaving a bare core. The size of the dust grains is known from the scattering of light. Most dust grains are in the region of one micron ($1 \mu m$) in diameter, although some grains are considerably smaller. The direction of the scattering of light (polarization) shows that the dust grains are elongated in shape. Interstellar dust grains have a high albedo for visual light which indicates a composition of silicates or water ices and organic polymers. If the grains were composed of graphite they would absorb light and appear black with low albedo.

The composition of interstellar dust varies in different types of molecular cloud. The dust grains in diffuse molecular clouds show a silicate core and some carbonaceous particles associated with a thin ice mantle and a simple chemistry of organic molecules. In a dense molecular cloud, on the other hand, the ice mantle is the dominant part of the dust grain and it contains a rich variety of organic molecules (Shaw 2006, pages 140-142).

4. Organic molecules in space

A large diversity of molecules have been identified in molecular clouds in addition to H_2. Diatomic molecules are the easiest to identify. Carbon monoxide (CO) was identified as the most abundant molecule after hydrogen early on in radioastronomy. Polyatomic molecules require telescopes with higher resolution for identification, and confidence in correct identification has not always been very strong. This has led to debates as to whether certain complex organic molecules exist in interstellar clouds or not. High-resolution rotational spectroscopy has, however, confirmed the existence of 120 different organic molecules in interstellar clouds. The largest polyatomic molecule identified so far is $HC_{10}CN$ with 12 heavy atoms.

Complex molecules occur in the core of molecular clouds in association with dust grains. The density of this part of the cloud can even be as high as a hundred million molecules per cubic centimetre (or 10^{14} molecules/m^3) (Zeilik 2002, page 320). Polyatomic molecules are thought to be preserved by the cold conditions within molecular clouds. Molecules can, however, exist up to a temperature of 4000 K (Shaw 2006, page 70).

Ethene (C_2H_4), an alkene; and benzene (C_6H_6), an aromatic carbon ring are present in molecular clouds, as well as alcohols, aldehydes and ketones. Carboxylic acids and amino acids were thought to be present, but this is now contended as a misidentification (Shaw 2006, page 118). Interstellar dust is composed of silicates (SiO_4) and ice. The ice is mainly water (H_2O), but also methanol (CH_3OH), ammonia (NH_3), carbon monoxide (CO) and carbon dioxide (CO_2).

Oxygen (O_2) is not found in space; however, all the gases found in anaerobic atmospheres are present: carbon dioxide, methane, water, hydrogen sulphide, ammonia, and also nitrogen. Klemperer (2008, page 378) writes that the non-polar species N_2 is observed by its polar protonated form HN_2^+. HN_2^+ is an abundant molecular ion detected in many regions of the galaxy and in other galaxies (Solomon et al. 1992). It provides a means of estimating the abundance of its non-polar parent N_2, which is the dominant form of nitrogen in molecular clouds.

Different clouds and different parts of clouds contain distinct molecular inventories. The Orion molecular cloud contains several saturated species such as ethanol (CH_3CH_2OH) and its CN analogue (CH_3CH_2CN), the simplest carboxylic acid (CH_3COOH, acetic acid) and methylamine (CH_3NH_2) (Shaw 2006, page 118). Water vapour is abundant in the heated region of Orion. Of particular interest is the three orders of magnitude increase of hydrogen sulphide (H_2S) in Orion's hot core.

The galactic centre

Radio astronomers have detected the presence of many very dense clouds, representing a considerable mass at the centre of galaxies. Some of these clouds form a nearly complete ring around the galactic centre. The densest of the clouds contain as many as 100 million particles per cubic centimetre. Young, hot stars have formed within some of the clouds ionizing the surrounding hydrogen. There is also a great abundance of cool stars in the galactic centre. Overall metallicity – the abundance of heavy elements seems to grow towards the galactic centre. Molecules such as ammonia (NH_3) and water vapour, as well as complex molecules such as ethanol (CH_3CH_2OH) have been detected within molecular clouds at the galactic centre.

11.1 LIST OF INTERSTELLAR MOLECULES

2 atoms:
H_2 Hydrogen
OH   Hydroxyl radical
CO   Carbon monoxide
SiO   Silicon monoxide
NaCl   Sodium chloride (salt)
KCl   Potassium chloride
CN   Cyanogen
CH   Methylidyne

3 atoms:
H_2O Water
HCN   Hydrogen cyanide
H_2S Hydrogen sulphide
SO_2 Sulphur dioxide
N_2H^+ Nitrogen ion
MgCN   Magnesium cyanide
CO_2 Carbon dioxide

4 atoms:
NH_3 Ammonia
H_2CO Formaldehyde
HC_2H Acetylene
SiC_3 Rhomboidal polar ring

5 atoms:
CH_4 Methane
HCOOH   Formic acid
HC_3N Cyanoacetylene

6 atoms:
CH_3OH Methyl alcohol
CH_3CN Methyl cyanide

7 atoms:
CH_3NH_2 Methylamine
CH_3C_2H Methylacetylene

8 atoms:
$HCOOCH_3$ Methyl formate (acetate)
CH_3C_3N Methyl cyanoacetylene

9 atoms:
CH_3CH_2OH Ethyl alcohol
CH_3CH_2CN Ethyl cyanide

(This list was taken from Zeilik 2002, page 310; Shaw 2006, page 117 (taken from Erhenfreund 2000) and page 147 (taken from Smith 2004); Hoyle & Wickramasinghe 1978, page 75; and Klemperer 2008, page 368).

5. The components of life

The principal complex molecules of life on Earth are sugars, DNA and RNA bases and amino acids which form proteins.

This is a list of the elements forming macromolecules within living cells:

- Sugars (ribose, deoxyribose and polysaccharides) are made from carbon, hydrogen, and oxygen with a hydroxyl group (OH).

- Energy-storing nucleotides (ATP) are composed of carbon, hydrogen and oxygen as a ribose sugar and a base with several phosphate groups (PO_4).

- The nucleotide bases adenine (A) and guanine (G) (purines), and thymine (T), cytosine (C) and uracil (U) (pyrimidines) are made from carbon, hydrogen, oxygen and nitrogen and nitrogen groups (NH, NH_2). DNA and RNA are made of nucleotides composed of a sugar (ribose or deoxyribose), a phosphate group and a nitrogenous nucleotide base.

- Amino acids are made of carbon, hydrogen, oxygen and nitrogen with an amino group (NH_3^+) and a carboxyl group (COOH), and sometimes a sulfhydryl group (SH). They have varying side-chains.

- Lipids (fat and oil) consist of a long hydrocarbon chain of carbon and hydrogen attached to a carboxyl group (COOH). Steroid lipids have fused carbon rings. Phospholipids have in addition a phosphate group (PO_4).

Twelve elements are present in the molecules discovered in interstellar molecular clouds: carbon (C), hydrogen (H), oxygen (O), nitrogen (N), phosphorus (P), sulphur (S), magnesium (Mg), sodium (Na), chlorine (Cl), fluorine (F), aluminium (Al) and silicon (Si). In addition there is ionized iron (Fe^{2+}), calcium (Ca^{2+}), potassium (K^+) and sodium (Na^+) in the interstellar medium. The CHON elements are the most abundant (Shaw 2006, page 117).

The most abundant atoms in the organic molecules of living organisms on Earth are CHON carbon, hydrogen, oxygen and nitrogen. Phosphorus and sulphur are also abundant.

Williams & Fausto da Silva (1985, 2008) list as the necessary elements for life in addition to CHONPS; magnesium (Mg), sodium (Na), chlorine (Cl), potassium (K), and calcium (Ca) as bulk biological elements. These are all found in the molecules and ions of interstellar clouds.

Trace elements necessary for bacteria, plant and animal life are: iron (Fe), copper (Cu), manganese (Mn), zinc (Zn), nickel (Ni), cobalt (Co), molybdenum (Mo), chromium (Cr), vanadium (V) and tungsten (W). These are all transition metals which generally play a role in enzymes. Silicon (Si), selenium (Se), fluorine (F) and iodine (I) are also mentioned. Most of these are known to exist in interstellar clouds.

The coincidence in the elements of which life is composed and those found in interstellar molecular clouds is remarkable; it is even more remarkable that the abundance of the elements found in space reflects their abundance as the components of life.

Crick (1981) in *Life Itself: Its Origin and Nature* points out that hydrogen cyanide (HCN) and formaldehyde (H_2CO) are reactive molecules that occur in interstellar clouds and that sugars, bases and amino acids would be easily synthesized from some of these molecules which occur in space.

Michael Denton (1998, page 78) writes,

"......... recent studies have revealed that interstellar space is filled with vast quantities of water, methane, ammonia, carbon monoxide, and many other organic compounds. The quantities are immense –up to 200 million solar masses in our own galaxy. The cosmos is literally overflowing with the basic constituents of carbon-based life."

6. Temperature and life

The temperature of interstellar clouds is a very important issue if we wish to determine if life could exist within these clouds.

Up to now, it has been thought that for life to exist on a planet orbiting a star, the planet must be located within the habitable zone in terms of its distance from the star. The habitable zone is defined as the zone where liquid water remains liquid. On the Earth's surface water freezes into ice below 273 K (0 °C) and boils into vapour at 373 K (100 °C).

It has always been claimed that Earth has life because it orbits the Sun within the habitable zone. In fact, measurements of the albedo of the Earth predict a surface temperature of 256 K or −17 °C. The warmer than predicted surface temperature of the Earth of 288 K or 15 °C is now attributed to the effects of global warming produced by water vapour and carbon dioxide in the atmosphere (Shaw 2006, page 203).

Textbooks generally state that interstellar clouds have an average temperature of 70 K, while molecular clouds may have a core temperature as low as 10 K. For example, Zeilik (2002, page 315) states that the infrared emission from cold dust in the core of the Orion Nebula shows that it has a temperature of 70K. Dark clouds typically have temperatures of 10 K (Zeilik 2002, page 320). 70 K is −203 °C and 10 K is −263 °C: these temperatures are much too low for life to be possible. However, with the knowledge that stars such as the Sun have surface temperatures of 5000 K and ionized hydrogen can reach 10 000 K, I formed the working hypothesis that in between these extremes of cold and heat, there must be zones of intermediate temperature that would be conducive to life. I later found the information I was looking for.

Within dark clouds there are often near-infrared sources which are unseen protostars enveloped in cocoons of molecules and dust. The dust absorbs the visual and ultraviolet light emitted by the protostar star, extinguishing it almost completely. The radiation absorbed has the effect of heating the dust grains to several hundred Kelvin which causes them to emit infrared radiation.

Zeilik (2002, page 320) writes,

".... as the protostar forms, the interior dust reaches about 300 K and so emits with a peak at 10 μm."

300 K is equal to 27 °C which is the temperature of a warm summer day. This is an ideal temperature for life to flourish.

On page 317, Zeilik writes that interstellar dust is often found surrounding cool stars which have surface temperatures of around 3000 K as circumstellar clouds. The classification of stars will be given in Chapter 13 and this will make things clearer. Interstellar dust is associated with protostars i.e. stars that are being born; cool red giants i.e. stars that are dying; and with failed stars known as brown dwarfs. To summarize we can say that interstellar dust surrounds these cool stars, but when stars become main sequence stars and their temperature increases, interstellar dust is dispersed and disappears.

The shells of dust that surround stars as circumstellar clouds have temperatures ranging from 2000 to 20 K. Consequently, there are certainly habitable zones in space especially around young and old stars.

Low pressure
The pressure in the low density conditions of space is very low compared to the pressure found on planets surrounded by atmospheres. In space the pressure drops to almost zero.

Lower pressure means that water remains liquid at lower temperature. The snow line of our solar system – the zone where water is only found as ice – is at a temperature of 160 K or −113 °C and a distance of 5 AU from the Sun. This is around the orbit of Jupiter.

Low pressure may extend the habitable zone by allowing water to remain liquid at a lower temperature. There may be forms of life adapted to the low pressure conditions of space which live at very low temperatures due to liquid water being available.

Extremophiles

Most bacteria prefer temperatures between 20 and 45 °C, but the temperature range for organic chemistry in extremophile bacteria is wider than for most other organisms. It extends between −20 °C and 110 °C. Extremophiles have been found growing inside glacier ice at one extreme and in hot springs at the other extreme. Bacterial growth is just possible at −20 °C (253 K) which is the temperature at which intracellular ice is formed.

Archaea and Eubacterial cells can adapt themselves to extreme cold by having a considerable salt content that depresses the freezing point of water. Bacteria living within glaciers in Iceland have also been found to produce proteins which act as anti-freeze, causing the ice to melt and form tiny pockets of liquid water.

Therefore, it appears that there are habitable zones within interstellar dust clouds with temperatures between 273 and 373 K (0-100 °C). The habitable zone can be extended by various other factors.

7. Life created throughout the universe

The association between interstellar dust, water ice and organic molecules in dark nebulas; the abundance of elements in space closely matching the abundance of elements composing life on Earth; and the existence of temperate zones within interstellar dust clouds allows the belief that there may be vast quantities of some sort of unicellular life in space.

I propose that God created unicellular life forms in vast quantities within the galaxies of the early universe. These life forms were closely associated with silicate dust, the origin of which will be discussed in Chapter 13.

The on-going process of the birth of stars as protostars in molecular clouds has provided a continuing source of light and warmth. Rings of dust may have provided anchorage for unicellular life forms in the low-gravity conditions of space. The metabolic processes of these interstellar life forms could have produced abundant water now found as ice, as well as, gases and complex carbon-based molecules.

The evolving nature of galaxies and stars would have produced conditions sometimes conducive to life and sometimes causing large quantities of life to perish. When life perishes, the elements of which it is composed would be returned to the interstellar medium, providing the raw material for life when it flourishes again under new circumstances.

When microscopic organisms die, the complex carbon molecules of living cells become dissociated into simpler carbon molecules. The result of this may explain why there are many different types of carbon molecule found in molecular clouds.

When conditions are right and elements or simple molecules are available as resources, unicellular life may increase exponentially and spread to new areas of the galaxy. Hence, interstellar clouds would undergo a cycling process with phases conducive to life and phases which banish all life. The cycle would be partly dictated by the evolution of stars through their different phases.

According to this idea, unicellular life was created on a massive scale within each galaxy. In the early universe, galaxies were much closer together than they are now. Could there have been some mixing between galaxies, or did each system remain in isolation? When a large galaxy draws in a smaller galaxy, there would certainly be mixing of any forms of life carried in the interstellar clouds.

Candidates for life in space
Which are the likely candidates for interstellar life among the known forms of unicellular life on Earth?

Methanogens
The methanogens are anaerobic Archaea that practice chemosynthesis. They obtain carbon to form organic molecules from carbon dioxide in the atmosphere, and they produce energy by converting carbon dioxide and hydrogen into methane and water:

$$4H_2 + CO_2 \rightarrow CH_4 + 2H_2O$$

Methanogens can also live on organic matter converting it to methane and water. The gases involved in this chemolithotrophic mode of life are all abundant in molecular clouds and this type of metabolism does not require light, thus theoretically it could be practiced in the interior of a dark cloud or at a considerable distance from a star. On Earth methanogens are found living in very hot environments such as hydrothermal vents, as well as in very cold environments such as within glacial ice, thus the temperature range tolerated by these organisms is very wide.

Green and Purple bacteria

Interstellar clouds are illuminated by starlight, thus one would expect to find a form of life taking advantage of this energy by practicing photosynthesis. Green and Purple Eubacteria practice anoxygenic photosynthesis in anaerobic conditions on Earth. Green and Purple bacteria use bacteriochlorophyll for photosynthesis. This pigment absorbs longer wavelength light (nearing the infrared end of the spectrum). This type of radiation is produced by protostars and red giant stars.

The decomposition of organic matter in anaerobic conditions produces hydrogen sulphide, ammonia, fatty acids and amines. Green and Purple sulphur bacteria use hydrogen sulphide and light energy to obtain carbon from carbon dioxide to make cell material and water, and they deposit sulphur granules. They can also live on hydrogen gas instead of sulphide. Some can live on organic matter as heterotrophs in darkness.

$$2\ H_2S + CO_2 + light\ \rightarrow\ CH_2O + 2\ S + H_2O$$

Green and Purple nonsulphur bacteria live on alcohol or acids such as lactate or pyruvate. Some continue to live by obtaining energy from the anaerobic fermentation of organic matter when in darkness.

Hydrogen sulphide, ethanol, pyruvate (C_3) and carbon dioxide are found in interstellar clouds as the possible starting point of this type of metabolism. Anoxygenic photosynthesis does not produce oxygen as an end-product, but water, and water is found in large quantities in space.

Cyanobacteria

Cyanobacteria practice aerobic photosynthesis to obtain energy and fix carbon dioxide. In the anaerobic conditions of the early Earth, cyanobacteria produced the first free molecular oxygen that eventually made Earth's atmosphere aerobic. During the first aeons, the oxygen released by photosynthetic bacteria combined with reduced metals, especially iron, producing banded iron formations of oxidized iron. This meant that the atmosphere continued to be anaerobic, lacking in molecular oxygen. I make this point since, if cyanobacteria existed in interstellar clouds, the oxygen released may be used to produce silicates – the main component of dust, or it may be absorbed by metals rather than released as molecular oxygen.

Cyanobacteria require water for their metabolism since oxygen is released from the splitting of water. Under conditions of low light intensity cyanobacteria may, however, resort to anoxygenic photosynthesis using hydrogen sulphide as an electron donor, producing water as an end product. Many species of cyanobacteria fix nitrogen (nitrogen is available in interstellar clouds) and this makes them highly adapted to pioneering harsh conditions. Cyanobacteria are one of the most resistant organisms on Earth growing in all sorts of adverse conditions.

Cyanobacteria use chlorophyll a for photosynthesis. They also have pigments called phycobilisomes which absorb light of shorter wavelength in the ultraviolet part of the spectrum. Phycobilisome granules make cyanobacteria able to cope with radiation of harmful intensity.

Heterotrophic bacteria

Heterotrophic bacteria live on organic matter by practicing anaerobic fermentation. Various types of bacteria produce acids and acetate (CH_3COO^-) through fermentation. On Earth the fungus, yeast produces alcohol through the fermentation of organic matter. Alcohol (ethanol) has been identified as one of the complex molecules of space. Could it be the by-product of the anaerobic metabolism of living cells in space?

Thus, it is possible to envisage a community of single cell life forms living in molecular clouds in association with silicate dust metabolizing gases, using the light energy from stars and feeding upon the organic matter provided by each other, whether living or dead. These living cells would be the source of complex organic molecules in interstellar clouds.

The properties of bacteria

The properties of bacteria that would adapt them to life on an early planet or even life in space have been noted by various authors. Shaw (2006) notes some of the properties of extremophile bacteria in the context of the new science of astrobiology. Francis Crick (1981) noted them in the context of his Theory of Directed Panspermia.

There is a type of bacteria called *Deinococcus radiodurans* which can withstand very high levels of ultraviolet light. The *Deinococci* have a DNA structure that is tolerant to radiation. They also have high concentrations of carotenoid pigments that may protect them from radiation and thick cell walls made from unique lipids. They survive extremes of heat and cold, and desiccation, and it has been shown that they are able to survive the vacuum of space. The *Deinococci* are aerobic, producing acid from a few sugars. If any *Deinococci* lived in space they would have to live in association with cyanobacteria that would provide oxygen for their aerobic metabolism. On Earth, *Deinococci* have been found eating nuclear waste inside nuclear reactors.

These are the general properties of bacteria that may enable them to survive in interstellar clouds:

- Many bacteria are anaerobic not requiring oxygen, while others are facultative anaerobes capable of respiration in the presence of oxygen and fermentation without oxygen.

- Bacteria are small in size. The average size of bacteria is 1.1-1.5 microns wide and 2-6 microns (µm) long (Mader 2007, page 62). *Escherichia coli* are about 1 µm wide and 2 µm long. The Archaea are often even smaller. Bacteria are found on dust particles or in water droplets suspended high in Earth's atmosphere.

- Bacteria have rigid cell walls made of peptidoglycan that protect them from the outside environment. The cell wall of Archaea contains polysaccharides and proteins. Bacteria can usually be 'freeze-dried' a process that first cools the bacteria then extracts the water from them with only minimal damage to the cellular structures (Crick 1981, page 127).

- Bacteria make the full set of amino acids from any source of carbon. (Humans only manufacture half the set of amino acids necessary, the rest must be acquired in foods).

- Bacteria generally reproduce by simple division. Because sexual reproduction is not essential, bacteria are not compelled to search for a mate –this enables them to live in low density conditions. A bacterial colony can grow from a single individual.

- When bacteria are stressed they often produce spores. Bacterial spores from Earth have been shown to survive for years in space on spaceship equipment. The endospore is a specialized structure that resists heat, and thus enables bacteria to survive extremes of heat, and then multiply again after long periods of time.

8. Photochemistry

There are various theories about how molecular hydrogen and organic carbon-based molecules could originate in interstellar clouds. The production of molecular hydrogen which is the main component of molecular clouds will be discussed in Chapter 13 in connection with star birth. In accord with current theory, I consider that the conversion of atomic hydrogen into molecular hydrogen must involve natural physical processes.

In this section of Chapter 11 I will present some of the current views as to the origin of organic molecules in interstellar clouds. Current theory considers only abiotic hypothetical pathways for the production of complex organic molecules in space. In opposition to current theory, I have proposed that these complex molecules are produced by life involving metabolism and enzymes.

It is well understood that the production of molecules out of atoms in gas-phase is not a viable proposition since the atoms in interstellar clouds are too far apart for any realistic prospect of chemistry taking place. Therefore, the role of dust grains within a molecular cloud is thought to hold the key to understanding molecule formation. This theme will be elaborated on in Chapter 13 regarding the formation of molecular hydrogen. Regarding the formation of carbon molecules, it has been suggested that simple carbon molecules in dirty ice surrounding dust grains may be processed into more complex organic molecules by the energy provided by ultraviolet radiation.

Equilibrium thermodynamic models

The models used for this kind of photochemistry follow equilibrium arguments. These are kinetic models of chemistry describing reaction dynamics. Klemperer (2008) points out that these equilibrium thermodynamic models for the production of organic molecules are not borne out by the evidence. Klemperer states that in some instances high-energy isomeric forms of chemical species are quite abundant in interstellar clouds. A thermodynamic model would not predict this since it would be expected that a low-energy configuration would be produced in favour of a high-energy configuration. The example he gives is of HCN (hydrogen cyanide) and its high-energy isomer HNC being found in almost equal amounts in cold molecular clouds. He concludes that equilibrium models do not predict the observed molecular abundances in molecular clouds (Klemperer 2008, page 373).

Klemperer believes that cosmic ray ionization of helium produces carbon ions and these enter a gas-phase chemistry from which organic molecules are synthesized. (Klemperer 2008, pages 379-380).

The currently proposed models for producing organic molecules in space only involve natural physical processes; they do not involve life. The proposed sources of energy for this photochemistry are various forms of ionizing radiation and ultraviolet radiation.

It is important to point out that actual observation shows that extreme photon fluxes such as ultraviolet radiation have the general effect of breaking chemical bonds, preventing molecule formation and ionizing atoms by removing electrons. Strong radiation produces photolysis products which are the broken pieces of complex molecules i.e. complex carbon molecules are broken into smaller carbon molecules. However, some of the photolysis products are reactive chemical species which could join up to form larger molecules.

Since strong radiation breaks up molecules, it is difficult to envisage how strong radiation would also build up complex molecules.

Enzymes

The enzymes of life build up complex organic molecules, as well as releasing molecular gases and water.

Enzymes are the biocatalysts of metabolism. Enzymes orient molecules such that they can react together. The specificity of enzymes is due to the shape of their active sites. The chemical reactions catalysed by enzymes include electron transfers, group transfers such as hydrolyses, isomerizations and reversible molecular splittings. Enzymes build proteins and polysaccharides from their constituents; they unzip DNA for replication; and they break up complex molecules such as proteins into amino acids in digestion and in cell recycling.

When enzymes cause atoms to form a new chemical bond, they also remove the energy released. If the energy is not removed, the atoms of the newly formed molecular subcomponent will split apart again. Removal of energy is the reason why enzyme-mediated biochemistry works, and why photochemistry is unlikely to work. The problem of energy-removal has not been dealt with adequately in hypotheses proposing the thermodynamic production of organic molecules in space.

It is my belief that the complex organic carbon-based molecules found in molecular clouds have been produced by enzymes acting as part of the metabolism of cells. Simple carbon molecules may be the subsequent break-down products from complex molecules.

9. Exotic life and the structure of interstellar dust grains

In this section I'm going to take another look at interstellar dust grains and suggest how they might be associated with forms of life.

Polarization

The light from distant stars is often observed to be partially plane-polarized. This means that light waves show a preference for a particular direction of vibration. The interstellar dust grains that are scattering the light must be aligned to produce this polarization effect. It has been noted that the long axes of the dust grains are aligned perpendicular to the direction of the galactic magnetic field. This implies that the dust grains are not made of metals or graphite. A paper by Piirola et al. (1992) of the Vatican Observatory suggests that the orientation of interstellar dust grains (of which they found two sizes) depends upon the direction of light coming from a star (rather than the galactic magnetic field).

Piirola et al. (1992) write that optical densities indicate that the population of dust particles around the star V633 Cas has a considerable component of small particles, less than 1 μm (micron) across, while the thin circumstellar disk around the star β Pictoris has much larger particles over 1 μm across. The polarization pattern of dust around V633 Cas is centrosymmetrically orientated, except in the innermost area where the polarization lines are parallel. This suggests the presence of a circumstellar disk around the central source.

The orientation of dust grains around a star and the polarization of light lead one to entertain the idea that the light could be serving a purpose. (This is my opinion not that of the authors of the article). Light would be serving a purpose if interstellar dust grains were associated with a form of life practicing photosynthesis or using absorbed light as a source of heat necessary for metabolic processes. Green and Purple sulphur and nonsulphur bacteria practice anoxygenic photosynthesis, while methanogens are chemolithotrophs which do not require light but they do need a source of heat for their metabolism to function.

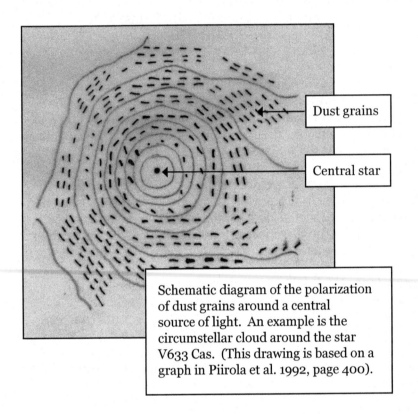

Dust grains

Central star

Schematic diagram of the polarization of dust grains around a central source of light. An example is the circumstellar cloud around the star V633 Cas. (This drawing is based on a graph in Piirola et al. 1992, page 400).

Miniature solar panels

As will be described in Chapter 13, the silicates of which interstellar dust grains are composed may have a totally separate origin; however, these silicate particles may be of use to forms of life living in space.

In Zeilik (2002, page 238) there is an electron microscope photograph of a particle of dust gathered from a comet. I will discuss the theoretical connection between comets and interstellar dust clouds in a later section of Chapter 11 and again in Chapter 17 of Part III. Cometary tails show the characteristics of sunlight scattered from small solid particles about a micron in size which is about the same size as interstellar dust. The composition of cometary dust and interstellar dust is also apparently similar. The dust grain from a comet shown in the photograph is about 10 μm in size and shows a fragile, open structure.

This gave me the idea that interstellar dust grains could have a structure similar to that of comet dust. The structure shows small scales composed of silicates attached in an open arrangement to a central core. The scattering of light shows that interstellar dust comes mainly in two sizes: half a micron (500 nm or 0.5 μm) and ten nanometres (10 nm or 0.01 μm) in diameter. These sizes are very small. However, it is possible that interstellar dust occurs as conglomerates of particles with each silicate scale of the conglomerate reflecting photons of light independently. This model of an interstellar dust grain is rather different from the one given in text books.

I formed the idea that there could be a form of life interwoven in the conglomerate of dust and holding the whole structure together by filaments.

The silicates of interstellar dust with their open structure (if this were the case) would not serve a protective function as silicates do for the single-cell forms of life known as diatoms on Earth. (Diatoms are single-cell algae that float in the surface waters of oceans; they have beautiful silicate tests as protective coverings).

In a flight of imagination it occurred to me that the silicate particles of interstellar dust could serve as miniature solar panels which a filamentous form of life would orientate to catch the light of stars. Mini solar panels could serve several purposes for life in space. They are as follows:

1. Black-coloured graphite-covered or tar-covered panels could absorb photons of light and thereby convey heat to a form of life living in the cold conditions of space. Interstellar dust clouds do absorb light energy from stars and the dust grains reemit infrared heat.

2. Light-coloured silicate panels could act as light reflectors so that as light is reflected among dust grains in the cloud, the light is effectively trapped. The reflection of light down-grades it from ultraviolet to visible wavelengths which would make it available for photosynthesis. Both the scattering and absorption of light is observed in interstellar dust clouds as extinction. It is possible that the light energy which does not

exit dark clouds is being transformed into chemical energy by anoxygenic photosynthesis.

3. It is known that dark clouds are not penetrated by UV radiation due to the deflection of radiation by solid dust particles. A form of life living within a dusty molecular cloud would be protected from harmful UV radiation by silicate panels reflecting UV radiation away.

4. Silicate panels could also serve as water-trapping surfaces. This would allow thin films of liquid water to be preserved between particles and made available to life. On Earth a type of rock called mica consists of flakes or sheets of silicates with a film of water and organic molecules in the gap in between each sheet. Interstellar silicate flakes could act in a similar way to foliated hydrous sheets of mica. It is known that water can remain liquid at temperatures well below freezing point as thin films on the surface of rock and ice crystals. A film of water only a few microns thick may provide enough water for a bacterium to survive.

5. The many panels of a conglomerate would provide a large surface area for the trapping of gas molecules within a molecular cloud. These molecules would be necessary to metabolism.

6. Conglomerates of silicate particles would form an anchorage for a unicellular form of life in space. Bacteria on Earth use silicates as anchorage. For example, oil spills can be cleaned up by adding bacteria that eat petroleum products along with powdered white clay to form islands in the oil for the bacteria to cling to (Canby 1993). The many faceted dust conglomerates would also have ice welding the whole structure together, formed from water produced by anaerobic metabolic processes.

Light mill effects

Interstellar dust clouds exhibit movement and circumstellar dust clouds rotate around the central star. It is possible that this movement is produced by a radiometer effect.

A radiometer also known as a light mill is a contraption with four vanes spinning inside a sealed bulb with a vacuum. I remember a light mill at my great aunts' house that would spin when the sun came out. The vanes are black on one side and white on the other. The light mill works by sunlight warming the dark side of each vane more than the white side. The vacuum means that the bulb contains only low density air molecules which bounce off the black side of the vane with greater velocity because it is hotter, than off the cooler white side. This causes the light mill to spin in the direction away from the dark side of the vanes. The force provided by collisions of molecules in a rarefied gas is much greater than the force provided by photons from sunlight. Sunlight reflects off the white side of the vanes, but does not provide sufficient force to turn the light mill in the opposite direction.

There is the possibility that miniature silicate panels in interstellar dust grains act as light mill vanes by being light in colour on one side and coated in black tar (kerogen for example), on the other. Space has low-density vacuum conditions. There would be enough molecules in a molecular cloud to cause the dust grains to collectively move in a direction away from the dark side of the particles. This could give momentum to the circumstellar cloud.

A filamentous life form

Movement may be necessary for life in space since any life form would have to encounter gases such as CO_2 and CO in the molecular cloud as a metabolic food source.

On Earth, microaerophilic bacteria such as *Beggiatoa* and *Thiothrix* capture and oxidize hydrogen sulphide from the air to generate energy. Their filamentous form helps them to capture the gases they need. One would expect a bacterial form of life adapted to the low-gravity and low-density conditions of molecular clouds to be filamentous in form. This would aid in trapping the necessary gas molecules from the cloud. Movement of the dust particles in the interstellar cloud would also be indispensible to life.

If interstellar life were often filamentous in form, this would have implications for its size in comparison with bacteria on Earth. Bacterial cells are much smaller than eukaryotic cells. Bacteria vary in size from 0.1 microns (0.1 μm) up to 500 microns (500 μm) in length (Atlas 1981, page 23). The average size of bacteria is 1.1-1.5 microns wide and 2-6 microns long (Mader 2007, page 62). Cell inclusions are measured in nanometers (nm).

Archaea are of a similar size to Eubacteria, but they come in more diverse shapes than Eubacteria. Shapes include spheres, rods, spirals and plates. Archaea of the *Thermofilum* species has needle-like filaments of less than half a micron in diameter (Krieg 2005). It is a form like this that one would expect to find in space.

It is known that interstellar dust grains have a large component of water containing organic molecules. It is possible that this forms a gel secreted by a filamentous form of life. On Earth cyanobacteria secrete a gel which has the protective function of screening out UV radiation that would cause damage to the bacterial cells.

Archaea abound in extreme habitats and there are also many extremophile bacteria. Bacteria are found living inside rocks below the Earth's surface that can withstand high pressure. Some bacteria live beneath the permafrost of Siberia and within glaciers in Greenland that can withstand low temperature. Other microbes living in hot, dry deserts and in near-boiling springs can withstand high temperatures. Bacteria are found floating in the air adhering to particles of dust. This shows us that bacteria on Earth have adapted to very wide-ranging conditions. However, we are not likely to find any exotic bacteria on Earth adapted to the conditions of interstellar molecular clouds i.e. conditions of low-pressure and low-gravity, since these conditions do not exist on Earth. We can only note the ability of bacteria to adapt to extraordinarily harsh conditions and thrive.

What we do know is that when bacteria encounter favourable conditions they grow exponentially. A dust cloud heated to 300 K (27 °C) by a central star or a cool dark cloud able to trap energy from starlight by silicate flakes could present ideal conditions to an interstellar form of life.

It is possible that interstellar life could proliferate and bloom creating enormous molecular and dust clouds in the right conditions. A microbial community could form composed of species with complementary requirements that allow them to avoid too much direct competition. This is observed in microbial communities on Earth; species live on the metabolic by-products of other species.

Liebig's law of the minimum states that microorganisms will continue to reproduce until an environmental or nutritional factor limits their growth. Limiting factors to microbial growth include availability of water, light energy, chemical compounds, nutrients and factors such as temperature, pressure, pH and salinity. Excess nutrients may allow unlimited growth of microorganisms, but rapid growth usually depletes nutrients and sometimes results in the build-up of toxic waste products that limit further growth.

Unicellular life forms may proliferate and bloom in molecular clouds on a scale not known on Earth.

11.3 TWO DIFFERENT CURRENT MODELS OF AN INTERSTELLAR DUST GRAIN

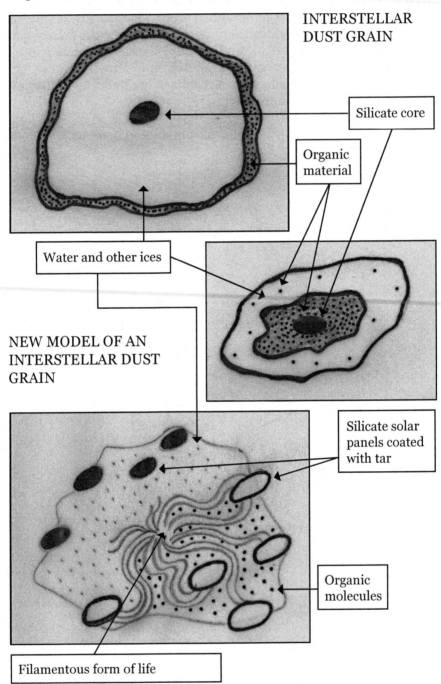

Silicate core

Organic material

Water and other ices

NEW MODEL OF AN INTERSTELLAR DUST GRAIN

Silicate solar panels coated with tar

Organic molecules

Filamentous form of life

10. Panspermia and other theories

There have been several theories based on the idea that life evolved elsewhere in the universe. For example, that life evolved on distant planets and travelled to the Earth.

Lord Kelvin (1871) was one of the first to propose that life may have come from space to the Earth, but the idea found greater elaboration with the Swedish physicist Arrhenius at the end of the 19th century. Arrhenius proposed that microorganisms ejected from the atmospheres of life-bearing planets elsewhere in the galaxy drifted through interstellar space as spores. Propelled by the pressure of light falling on them, some of these spores wafted in from space to the Earth and gave rise to life on Earth. Arrhenius called his idea Panspermia meaning 'seeds everywhere'.

Directed Panspermia
The idea of Panspermia was remodelled by Francis Crick (of DNA fame) and Leslie Orgel in an article published in *Icarus* in 1973 on the Theory of Directed Panspermia. This theory centres on the idea that the seeds of life were purposely spread by an advanced extraterrestrial civilization.

The idea presented in *Life Itself* by Crick (1981) starts with life evolving nine billion years ago on a distant planet. This extraterrestrial life develops into intelligent life who discover science and technology four to five billion years later. They decide to send life to other planets in the galaxy; however, the journey takes too long for multicellular organisms or themselves to be sent. They hit on the idea of sending packages of bacteria in an unmanned spaceship. The spaceship protects the microorganisms during the long journey of hundreds or thousands of years through space. When approaching planet Earth, delivery packets containing bacteria are able to withstand entry through the atmosphere and plunge into the Earth's primitive ocean where they burst open. The primitive ocean is full of all the ingredients you need to make life, and the microorganisms begin to multiply. Then life evolves again on our planet. In this way, Crick proposed that many sterile planets were seeded with life.

Crick discusses criticisms of Directed Panspermia in *Life Itself* on page 148. He relates that his wife voiced *more than once* that it is not a real theory, but merely a science fiction –I'm sure it does have appeal to those who enjoy science fiction, but it does not help us answer any scientific questions.

Life-bearing comets

Fred Hoyle and Chandra Wickramasinghe working on radioastronomy that aims to identify the molecules found in interstellar clouds took up a different line of thought. They describe their ideas in *Life Cloud* published in 1978.

Analysing infrared emission data from interstellar dust, Hoyle and Wickramasinghe concluded that the dust was composed of an organic material with the elements carbon, nitrogen and oxygen. They proposed that interstellar dust is made of cellulose or a related polysaccharide. Eventually, they found that an ensemble of polysaccharides and some simple hydrocarbons provided the best fit to a wide range of infrared data.

Gaseous formaldehyde is one of the most ubiquitous organic molecules found in interstellar space. They hypothesized that formaldehyde could have originated in the outflows of gaseous material from hot, young stars. They further hypothesized that formaldehyde could have formed polymers within the interstellar cloud due to a series of compressions and expansions of the cloud.

To answer the question; How did life start on Earth? –they write on page 128,

"If interstellar space is full of prebiotic molecules –the type or subset which consists of those molecular structures intimately associated with terrestrial life –it is almost self-evident that the origin of life on Earth merely involved a piecing together of interstellar prebiotics."

According to the view of Hoyle and Wickramasinghe, the vehicles which brought interstellar organic molecules to Earth were comets. They believe that comets brought water to Earth that would form the oceans; gases that would form an atmosphere; prebiotic organic chemicals that would fill the oceans; and finally life which had evolved extraterrestrially within comets. They explain on page 130 that heat released from chemical reactions between organic molecules could have melted a fraction of the underlying ice of the comet providing liquid water. If adequately insulated against heat loss by overlying frozen surface layers, this situation could have provided the best location for the emergence of life. All the building blocks of life could have been present in the melted ice – polynucleotides, polypeptides, porphyrins and carotenoids. These pre-existing macromolecules could, they believe, have evolved into anaerobic prokaryotic bacteria within comets.

Thus, according to the 'Life Cloud Theory', the emergence of life on Earth may have occurred 4000 million years ago by the arrival of a life-bearing comet making a soft landing into the already formed ocean. They write on page 157,

"The point of view we have developed so far implies that the essential biochemical requirements of life exist in very large quantities within the dense interstellar clouds of gas, the so-called molecular clouds. This material became deposited within the solar system, first in comet-type bodies, and then in the collisions of such bodies with the Earth. We might speak of the Earth as having become 'infected' with life-forming materials, and other planets moving around other stars would be similarly infected."

A further claim (which brought Hoyle and Wickramasinghe much criticism and probably lost Hoyle a Nobel Prize for his work in radioastronomy) was the claim that Darwinian evolution alone could not account for the vast elaboration of species as evidenced in the palaeontological record. Hoyle and Wickramasinghe argued that the gene elaboration involved in the great surges of evolution results from the effect of continuing viral invasions from outside. The proposal that unicellular life continues to enter the Earth's atmosphere and is responsible for epidemics, new diseases and the genetic novelty which drives forward macroevolution has not had a very good reception.

Carbon-bearing comets and meteorites

It is currently accepted that the carbon molecules that formed life on Earth came from space either in comets or in carbon-bearing meteorites. It is also recognized that carbon reaches the Earth as in-falling dust. The present rate of in-fall is 5000 tonnes per year. Some accept the view that life may exist elsewhere in the universe. Shaw (2006) suggests that there was a seeding of CHON chemistry on planets (CHON stands for carbon, hydrogen, oxygen and nitrogen elements). Concerning an extraterrestrial origin of life, Shaw tentatively mentions Panspermia –the theory that life was seeded everywhere somewhat randomly. He entertains the idea that life is formed wherever the chemistry will allow it, either randomly or directed by some guiding force. Having been formed everywhere; life would survive only where the planetary conditions are hospitable. Despite the adverse conditions in space for the survival of life that include low pressure, cosmic rays and intense UV radiation, bacterial spores could be transported to other planets protected inside meteorites. In support of this theory Shaw writes on page 9,

"Measurements of the survival of bacteria in space suggest that, in the form of spores or dried cells, the survival in space is possible at least for the 6 years that the experiments have been taking place. The transfer of life from planet to planet is then a real possibility. The recent extensive analysis of meteorite ALH84001 suggests that there are structures within the rock that look like fossil organisms."

11. Meteorites

It is now time to examine the nature of meteorites, comets and meteoroids in view of assessing their potential as carriers of unicellular life. Their origin, whether from within the solar system or from outside it is crucial in addressing this question.

The estimates vary widely, but Taylor (1998, pages 177-178) states that about 100 tons of meteoritic material falls on the Earth every day. Most of this material is dust, but larger meteorites weighing several kilograms also fall upon the Earth at a steady rate. Occasionally, much larger meteorites appear. The Allende meteorite sited over Mexico in 1969 had an original size of an automobile. It broke into thousands of smaller stones ranging in size from 1 gram to 110 kilos strewn over 50 km of desert terrain. The hundreds of meteorites collected equalled at least 2-3 tonnes of specimens[1].

---

[1] A metric tonne = 1000 kilograms, while a British ton = 2240 pounds or 907.185 kilograms.

Meteorites are classified into three types:

- Irons are 90 % iron and 9 % nickel, with traces of other elements. They are heavy and have a melted appearance having once been molten iron.

- Stones are composed of silicate materials. Many stony meteorites contain silicate spheres called chondrules embedded in a smooth matrix and they are known as chondrites.

- Stony-iron meteorites often have small pieces of stone set in iron.

There is a class of stony meteorites known as carbonaceous chondrites which contain a layer of organic material called kerogen between the chondrules. Kerogen gives the meteorite a dark colour. Carbonaceous chondrites contain significant amounts of water (about 10 %) and volatile materials (gases).

Hoyle & Wickramasinghe (1978) further classify the carbonaceous chondrites into three types:

- Type I carbonaceous chondrite: contain sulphur in the form of sulphates, iron as nickelian magnetite and silicates in an amorphous hydrated form. They have a fine-grain structure indicating a low temperature assembly of separate small particles. They have a total carbon content of 3.5 %.

- Type II carbonaceous chondrite: contain elementary sulphur, iron in the form of a hydrated silicate and 2.5 % carbon. They have micron-sized clumps consisting of many separate grains and they contain 'chondrules' or insertions of material measuring up to one centimetre across.

- Type III carbonaceous chondrite: have only 0.5 % carbon, less hydration and a less conspicuous chondrule structure.

The question has arisen: Does the organic material found in carbonaceous chondrites indicate the presence of life? Are the small structures found in these meteorites fossils of microbes?

Hoyle & Wickramasinghe (1978) report in *Life Cloud* on the findings of Claus & Nagy in 1963 on organized structures 4 to 30 microns in size found in carbonaceous chondrites.

"Five different types of fossil-like structures have been observed, of which four vaguely resemble certain types of single-celled terrestrial organisms living in water, such as algae. Differences in detail, however, make it difficult to identify the 'fossils' with any known terrestrial species. Electron micrographs of these 'fossils' show that most of them are roughly spherical, and there is evidence of a substructure resembling cell walls, as well as residues of cell nuclei pores and flagella-like formations. The fifth type of fossil was found to be totally unlike any known terrestrial organism. It is roughly hexagonal, with three of the bounding surfaces thicker than the others, and the entire hexagonal structure is surrounded by a spherical halo. In several cases the fossil structures observed by Claus and Nagy show constrictions in the central parts of elongated objects, and this suggests a process similar to cell division." (Hoyle & Wickramasinghe 1978, pages 110-111).

Hoyle and Wickramasinghe accepted a possible biological origin to the Claus-Nagy structures. They argued that primitive living organisms may have evolved in the mixture of organic molecules, ices and silicates in cometary nuclei as they became more concentrated with each passage of the comet by the Sun (Hoyle & Wickramasinghe 1978, page 111). They believed that Type I carbonaceous chondrites are spent comets.

There has been, however, increasing scepticism about the biological origin of the Claus-Nagy structures as there was no explanation as to how microbes could have become fossilized within meteoritic rock. An alternative explanation is that they are mineral grains which have acquired coatings of organic molecules by some non-biological process. Contamination by microorganisms on Earth has also been suggested.

Nevertheless, carbonaceous chondrites have continued to be studied in great detail. Some scientists such as Preston Cloud (1988) continued to be enthusiastic enough about the biological origin of meteorite organic material to write that carbonaceous chondrites have 50 amino acids and some have all five basic structural components of DNA and RNA suggesting the presence of life.

There are two carbonaceous chondrite meteorites that have been studied in fine detail; the Murchison meteorite found in Australia in

1969 and the meteorite ALH84001 found in Allan Hills, Antarctica in 1984.

The Murchison meteorite contains 2 % carbon as carbonates, soluble compounds such as amino acids and kerogen. The kerogen consists of aromatic and aliphatic macromolecules. There are hydrocarbons with chain lengths of C_{15} to C_{30} both straight and branched. There are polyaromatic hydrocarbons (PAHs) including pyrene and fluoranthrene, polar hydrocarbons such as aromatic ketones, alkyl and aryl ketones, nitrogen and suphur heterocycles; and most intriguingly purine and pyrimidine analogues.

Shaw writes,

"Hot aqueous extraction from the Murchison kerogen produces alcohols, ketones, mono- and dicarboxylic acids, amines and sulphonic and phosphonic acids –surely with such a suggestive mix of chemicals life is inevitable. Perhaps this is too good to be true." (Shaw 2006, page 172).

Shaw goes on to say that the traces of the Murchison and Allende meteorites when compared to the biological deposit known as fish clay found in Denmark show an almost identical spectrum graph from their carbon molecules (Shaw 2006, pages 173-174). If the decomposition products of fish clay are so similar to the decomposition products found in these meteorites, do they not all have a biological origin?

The meteorite ALH84001 contains carbonate globules thought to be deposited by water. Polyaromatic hydrocarbons (PAHs) are associated with the chondrules in the interior of the meteorite. The electron microscope shows tiny tube-like forms 500 nm in length on the globules of carbonate which have an uncanny resemblance to some type of nanobacteria. The excitement of this discovery caused NASA to found the subject of astrobiology. However, others believe the features to result only from molten material solidifying rapidly or from the precipitation of minerals into inorganic crystals (Shaw 2006, pages 177-178).

12. Comets

Comets are large dirty snowballs that enter the inner solar system at random approaching the Sun in elliptical orbits and swinging out again, to return at a later date. (The orbits of comets will be dealt with in Chapter 17. In this Chapter we will look at the composition of comets).

Most of the comets we see have an icy nucleus a few kilometres in size surrounded by a coma of vaporized gases and dust. As the comet approaches the Sun a tail forms of dust and gases expelled from the nucleus which may stretch for millions of kilometres. The tail often bifurcates into separate gas and dust tails. The gas tail contains ions which become caught up in the magnetic field lines of the Sun, while the dust and neutral molecule tail is generated by the motion of the comet.

Comets typically rotate every 6 hours. The temperature varies from 50 to 195 K (−223 °C to −78 °C) during the day-night rotation at perihelion (when it is approaching the Sun) and 50-160 K at aphelion (when it is away from the Sun).

The nucleus of a comet cannot be observed by telescope, but its composition is inferred from the composition of the surrounding coma. The gas and dust composition is studied using spectral analysis. Missions sent to fly close by Halley's Comet confirmed that comets are made principally from water ice and silicate dust. The ice evaporating as the comet approaches the Sun also contains other molecules such as carbon monoxide (CO), methane (CH_4), carbon dioxide (CO_2), methanol (CH_3OH), formaldehyde (H_2CO), hydrogen cyanide (HCN), ammonia (NH_3), hydrogen sulphide (H_2S) and hydrocarbons (C_2H_2), (C_2H_6), (C_3H_4).

Apart from gases such as carbon monoxide, carbon dioxide and molecular nitrogen, the gas tail contains radicals, ions and simple molecules including CN (cyanogen), CH (methylidyne), OH (hydroxyl radical), NH_2 (amino group), C_2, C_3, CO^+, CH^+, CO_2^+, N_2^+, OH^+, H_2O^+, and Ca^+ (calcium ion).

The fact that large molecules such as methanol, formaldehyde and more complex molecules have been detected in the coma of comets indicates that a greater diversity of molecules may be found below the surface of the nucleus. Shaw writes that simulations of cometary ice chemistry in the laboratory lead to an inventory of chemicals that may well be found below the ice surface. "The analysis of the laboratory ice experiments reveals a diverse array of biological-looking molecules with the potential to seed life." (Shaw 2006, page 185).

Shaw attributes the existence of complex organic molecules in comets to ultraviolet photon, cosmic ray and thermal processing on the comet's surface of simple molecules into a rich diversity of complex organic molecules (Shaw 2006, page 185). Hoyle and Wickramasinghe see things the other way round. They write that small molecules, radicals and ions observed in the coma of comets are unlikely to exist in the nucleus in this form for any appreciable length of time. The reason for this is that their chemical structures are inherently unstable. They believe that it is much more likely that the small radicals and ions are fragments of larger organic molecules which break up as they evaporate from the nucleus and are exposed to ultraviolet radiation from the Sun. There may be complex organic molecules in the nucleus of comets that have not been detected (Hoyle & Wickramasinghe 1978, page 103).

In the opinion of Hoyle and Wickramasinghe, many of the radicals observed in comets are dissociation products of organic polymers, such as polysaccharides. They hypothesize that a mixture of ices, organic polymers, silicate and graphite grains in the cometary nucleus become eroded by the action of solar radiation as the comet approaches the Sun. Hydrated polysaccharides could provide a single 'parent molecular species' for many of the observed radicals and molecules in the coma. This includes ionized water (H_2O^+), the free radicals OH, NH_2, CH, as well as molecules such as diatomic and triatomic carbon (C_2, C_3) which result from the breakup of pyran rings (Hoyle & Wickramasinghe 1978, page 104).

Halley's Comet
Halley's Comet passed by the Earth in November-December 1985 and in March-April 1986. Spacecraft fly-bys observed that the nucleus of Halley's Comet is peanut shaped and measures 15 km by 7 km. It is black and bumpy, and coated in tar-like material as a residue of organic compounds. The sunlit side has jets of vaporizing ice composed of 80 % water vapour and 20 % dust. The dust is mostly carbon in the form of graphite with a small amount of silicate material.

The jets spurt out 20 tons of material per second. Halley's Comet sheds a layer a few metres deep from its nucleus with each passage by the Sun. Comet Halley has spent around 23 000 years in the inner solar system and has orbited the Sun some 300 times in the opposite direction to the planets. As its material becomes exhausted, it will eventually disappear.

An article in Encyclopedia Britannica on Halley's Comet states that the nucleus of the comet is covered with a porous black crust. When measured, the crust was found to have a temperature of approximately 350 K or 80 °C. This temperature is much higher than the temperatures generally given for comets of well below zero centigrade. The article states that the unexpectedly high temperature is attributable to surface absorption of solar radiation.

I would make the comment that 80 °C is within the upper limit of temperatures tolerated by extremophile bacteria. Therefore, it is a temperature that makes life in comets possible. Secondly, this measurement of temperature on a black kerogen-coated surface should be retained in the memory when reflecting upon the possible conditions within an interstellar dust cloud where dust grains also possibly present this phenomenon.

Shooting stars
Meteoroids are particles of dust larger than 0.1 mm that enter the Earth's atmosphere and burn up while still high in the atmosphere. They produce what is commonly known as shooting stars. Micron-sized dust also enters the Earth's atmosphere, but it does not become seriously heated in transit, and drifts down to settle on the Earth's surface.

Shooting stars are seen at particular times in the year when the Earth is passing through accumulated trails of particle debris left by the passage of comets (Hoyle & Wickramasinghe 1978, page 108). Meteoroids (particles travelling through space) and meteors (particles that have entered the Earth's atmosphere and burn up) are thought to be of cometary origin. Larger sized pieces which hit the Earth as meteorites are thought to have a different origin.

The origin of meteorites and comets

It was proposed by Hoyle and Wickramasinghe in *Life Cloud* that meteorites and comets do not originate from somewhere within the solar system, but from outside it and have been closely associated with interstellar clouds. They state that, evidence that a component of meteorites is extrasolar is the isotopic ratio of neon gas which is anomalous with solar system values; this suggests an interstellar origin to some components of meteorites.

"In view of the strikingly close similarities of chemical composition between interstellar matter, comets and carbonaceous chondrites, we consider it far more likely that the carbonaceous matter in these meteorites represents a primordial interstellar component. In other words, the molecules of life have reached this planet through the addition of cometary and meteoritic materials, which –as we have discussed in detail in earlier chapters – themselves reflected the organic materials still pervading the interstellar clouds of gas and dust." (Hoyle & Wickramasinghe 1978, page 113).

It is also my opinion that meteorites and comets have an extrasolar origin. I believe that carbonaceous chondrite meteorites are bodies composed of silicates and iron that have journeyed through interstellar molecular clouds before entering the solar system. As these bodies moved through the denser parts of interstellar clouds they became coated with tar-like hydrocarbons (kerogen) as well as silicate dust. At this stage they may have formed part of a larger body which subsequently split into many smaller meteorite pieces. This explains the organic component of carbonaceous chondrites. The origin of the moving body will be explained in Chapter 17 of Part III.

I see comets as conglomerations of interstellar dust grains with a very high water ice component. The complex carbon molecules of comets would have been acquired along with the interstellar dust component when the nucleus of the comet was formed. The very low temperatures of comets may mean that they represent a deep-freeze version of interstellar life. These snow-balls several kilometres in size may have been rolled together by shock waves going through interstellar clouds from supernova explosions (comets typically rotate on their axes every 6 hours). The momentum generated would also have launched the snowball comets in different directions including towards our solar system. Upon orbit around the Sun, cometary material evaporates and the complex molecules become dissociated into ions, radicals and smaller molecules which surround the nucleus as a coma and trail behind as a tail.

The different composition of meteorites and comets reflects their different formation as well as their journey through different parts of the same interstellar cloud or different clouds.

The idea of an interstellar origin to comets differs from the current theory which locates the origin of comets as ice left over from the formation of the solar system found orbiting the Sun in the hypothetical Oort cloud and Kuiper belt. These theories will be discussed in more detail in Chapter 17 of Part III.

13. Planetary rings

The giant gaseous planets Jupiter, Saturn, Uranus and Neptune in the outer solar system are orbited by rings of material ranging in size from dust particles to small moons. Planetary rings are an example of dust in space. I believe that planetary rings may be connected to the bigger picture of interstellar dust clouds.

I will start with a description of the rings:

Rings of Jupiter The main ring lies at 125 000 km from Jupiter's centre and is 7000 km wide. Within this ring tenuous material reaches to the planet's atmosphere. An outer tenuous ring extends out to 225 000 km. The rings are thin with only about 30 km thickness. The scattering of light shows that the ring particles are only about 10 microns in diameter (Zeilik 2002, page 207).

Rings of Saturn There are three main rings, A, B and C which reach from 71 000 to 140 000 km from Saturn's centre. These very famous rings, although very wide, are very thin, being only a few kilometres in thickness. The A ring is relatively smooth, while the B and C rings are composed of hundreds of light and dark ringlets with widths as small as 2 km.

Saturn's rings are composed of lumps of ice and rock covered with ice. The ice in Saturn's rings reflects light and gives an albedo of 80%. The ice does not evaporate since the temperature of the particles is about 70 K (-203 °C). The average size of the particles is about one metre in diameter, although there is a whole range in sizes, up to boulders tens

of metres across. There is also fine dust, especially in the outer rings. The ethereal E ring may consist entirely of dust, and presents itself more like a cloud than a disk. The particles in the rings take about a day to orbit the planet.

Rings of Uranus There are at least eleven rings with moons embedded in the rings around Uranus. The smallest ring is only a few kilometres wide, while the largest is about 100 km wide. The rings are composed of particles a few metres across mixed with small particles only about a micron in size. The material that makes up Uranus' rings is extremely dark with an albedo of 5 %. This indicates that it is made of dark carbon materials.

Rings of Neptune There are four or five rings. The brighter outer rings extend to 53 000 km and 62 000 km with a tenuous ring in between them. There is an inner ring 2000 km wide. The outer ring is not uniform, but has three thicker regions.

The rings surrounding the gaseous planets all lie close to the planet – within three radii of the planet. The diverse composition of the rings is posing problems to scientists attempting to explain the origin of the rings. The clumping of the material in the rings of Neptune is not understood, nor are the sharp edges defining rings. A hypothesis of unseen shepherding satellites has been put forward to explain how the rings would be herded along and kept distinct from each other.

It would be expected that the material in the rings would be swept up by the central planet within a few hundred million years. The very small particles in the rings of one to ten microns diameter would spiral in towards the planet fairly quickly. So, the stability of the planetary rings cannot be explained. Taylor (1998, page 110) believes that the rings may be constantly replenished by the broken up remains of captured comets or the small-sized particles may come from tiny moons that have not yet been found due to their small size.

Planetary rings – an interstellar origin?

I believe that the planetary rings could be close up examples of interstellar dust clouds. Circumstellar dark clouds may, in fact, exist as rings. These are my reasons for linking planetary rings and interstellar dust clouds:

- The rings do all contain fine dust of a similar size to interstellar dust, although some of the rings contain much larger bodies too. The larger bodies range in size from pieces of a few centimetres to a metre to a few metres in diameter to the size of small moons. Larger pieces may also exist within interstellar clouds, but their presence cannot be detected at such great distances. This is because large pieces do not interact with light the way finely divided particles do, so spectral analysis will not reveal their presence.

- The composition of the planetary ring material is similar to that of interstellar dust – it is composed of silicates (rocky material), water in the form of ice and sometimes complex carbon molecules giving a dark coating.

- The ring structure of the dust orbiting Jovian planets may appear more complex because it is observed relatively close up within our own solar system instead of far away.

- The fixed distance range between the central source of heat and the ring of dust appears to depend on the amount of heat emitted. This suggests that the dust in the rings may have been associated with life utilizing the available heat at least at one time. The same would be true for interstellar clouds of dust surrounding protostars as for rings of dust surrounding planets in our solar system.

The question as to how rings of dust of interstellar origin could be found orbiting planets in the outer solar system will be answered in Chapter 17 of Part III. Chapter 13 of Part II deals with the identity of the gaseous planets which may, in fact, be very small failed stars known as brown dwarfs. In Chapter 17 the formation of the solar system will be discussed with the proposal of a new explanation for its origin.

There is the possibility that the ring systems are associated with forms of unicellular life living in association with the dust.

There is the greater possibility that the ring systems are the remnants of circumstellar dust clouds. The small amount of infrared energy emitted by the Jovian planets may be only enough to support relatively small and tenuous rings compared to the larger, denser dust clouds around newly forming protostars which emit greater amounts of infrared heat. Therefore, the ring systems may be a frozen reminder of relatively recent thriving dust communities, but ones that no longer multiply themselves as before.

I have presented planetary rings as a possible example of interstellar dust as it is seen closer to home and in greater detail.

14. Debris of life

I have proposed that life may be found in the dark, dusty parts of interstellar molecular clouds. There may be blooms of unicellular life around protostars that provide gentler forms of heat than fully evolved stars. The dust of dark clouds could offer some protection against damaging radiation. As these forms of life proliferate they would form molecules – gases, water and organic molecules found within dark clouds.

However, for the most part, interstellar clouds are not inhabited by life and the organic molecules found there are not found as the components of life. Interstellar organic molecules have been described as prebiotic molecules – the building blocks of life. I propose that they are the postbiotic debris of life.

Space is a dangerous place; there are many cosmic hazards for life there. When a protostar reaches main-sequence and starts up its thermonuclear furnace, it will emit lethal UV radiation. It will also develop a solar wind carrying accelerated particles. Supernova explosions of dying stars may also accelerate particles that can do lethal damage. Thus, space is inundated with high-energy photons and criss-crossed by accelerated particles. One class of lethal high velocity particle is the cosmic ray.

The result of UV radiation emitted by bright stars is that large parts of interstellar clouds become ionized. This creates regions of hot ionized hydrogen atoms (HII regions). If any life had once inhabited these areas as molecular clouds, it would have been wiped out and the molecules converted into dissociated charged atoms.

If an accelerated particle collides with a hydrogen atom it can convert it into deuterium (hydrogen with a proton and neutron). Deuterium can replace normal hydrogen atoms in molecules. If it replaces the hydrogen atoms in water, heavy water is the result (D_2O). If D becomes incorporated into organic molecules such as enzymes instead of H, the result is disastrous to life. The problem lies in the fact that the bond strength of D is stronger than that of H, and this prevents enzymes from fulfilling their normal function in metabolism. Thus, deuterium poisoning is another hazard for life in space.

UV radiation, accelerated particles, thermal heating and deuterated species would be the causes of ecological disasters on an astronomical scale and would effectively sterilize large parts of interstellar clouds. Therefore, I am not proclaiming life in space where only the remains of life can be found.

I believe that life does continue to proliferate in some dark corners of interstellar clouds, but most of what is found is the debris of life.

Breakdown products
Breakdown products are smaller molecules resulting from the breaking of chemical bonds by high energy radiation and accelerated particles. The subcomponents of known complex organic molecules are often found in interstellar clouds.

Short wave-length radiation breaks the C–C bond leading to the breakdown of complex carbon molecules. The strongest chemical bond is C≡ O so this may explain why there is so much carbon monoxide (CO) in space.

The photolysis of water gives the OH radical and hydrogen:

$$H_2O + h\upsilon \rightarrow OH + H$$

Cells contain proteins, nucleic acids, polysaccharides and lipids. The raw materials for the synthesis of these polymers are abundant in space. The smaller molecular subunits may also represent the debris of life which has failed to survive.

The following examples of molecules found in interstellar clouds are noted in the literature as prebiotic molecules, but they may represent breakdown products:

Methane (CH_4), water (H_2O), carbon monoxide (CO), carbon dioxide (CO_2), hydrogen sulphide (H_2S), ammonia (NH_3) and hydrogen cyanide (HCN) or derived compounds. These are the basic ingredients for the synthesis of the organic polymers of life (Williams & Fausto da Silva 2008, page 461).

Formaldehyde (H_2CO) – a basic molecular unit of sugars and polysaccharides is abundant and widespread in the galaxy (Orgel 2008). The membrane lipids of Archaea are made from polyprenyl derivatives. The precursors are isobutene and formaldehyde (Ourisson 2008, page 431).

The pyrimidines and purines of DNA are formed from starting materials such as hydrocyanic acid, cyanamide, cyanoacetylene (HC_3N) and water. Adenine is a simple pentamer of hydrogen cyanide (HCN).

Hydrogen cyanide (HCN) and aldehydes (CH_2O) found in interstellar space combine in the absence of oxygen to form amino acids, especially if ammonia (NH_3) is present to catalyse the process (Preston Cloud). Formic acid (HCOOH) and methanimine (H_2CHN) could react to give the simplest amino acid, glycine (NH_2CH_2COOH). Carboxylic acids and glycine have been found in the interstellar medium (ISM).

Porphyrins have been detected in interstellar clouds with the formula $MgC_{46}H_{30}N_6$. The union of four C_4N rings forms the central core of a porphyrin. Porphyrins are the precursors to the green pigments of chlorophyll and the red pigments of haemoglobin. The primary ingredient is hydrogen cyanide (HCN). Nitrogen-bearing rings are also required for nucleic acids (Hoyle & Wickramasinghe 1978, page 98).

Over-bonding

If an event happens involving both heat and pressure, the bonds of organic molecules may break and reform. Secondary multiple bonding produces thick sticky substances like tar or kerogen. Organic substances with 'over-bonding' are not the components of life, but the decay products of life.

It is possible that shock-waves from nova and supernova explosions moving through an interstellar cloud could produce this situation. Hoyle & Wickramasinghe (1978, page 76) write that tar-like sticky organic molecules coating the interstellar dust grains may cause the dust grains to stick together to produce grains 10 to 30 times their original size. This could be the seed for the formation of a comet nucleus.

Hoyle & Wickramasinghe also write that studies of the insoluble organic matter in carbonaceous chondrite meteorites show that it has a highly condensed coal-like aromatic structure, with an infrared spectrum not very different from that of cellulose. They believe that this matter represents a modified form of interstellar cellulose which has been coalified.

On Earth, kerogen found in sedimentary rocks such as shale, is a decay product formed from ancient freshwater and marine cyanobacteria and algae; terrestrial plant resins, cuticle, spores and pollen; and animal decomposition. Kerogen has a very high molecular weight making it mostly insoluble, though a soluble portion is known as bitumen. Upon heating kerogen releases first crude oil, then natural gas. Oils are heavy hydrocarbons, while gases are light hydrocarbons, and inert kerogen can be converted into graphite.

Chirality

Chirality refers to the asymmetric form of carbon molecules. Optical activity shows that carbon molecules can be 'left-handed' or 'right-handed'. The molecule rotates the plane of polarized light to the left (levo) or to the right (dextro). Life is composed only of L-form amino acids and D-form sugars (polysaccharides). The reason for this homochirality displayed in all life forms is that a mixture of chiral structures in a cell would cause them to inhibit each other. The binding site of an enzyme can only be accessed by the correctly shaped molecule, just as the right hand must go into a right-hand glove, and will not fit into a left-hand glove.

The fact that organic molecules produced by life always exhibit homochirality –L-form amino acids and D-form polysaccharides makes homochirality an important biomarker – an indication of the presence of life.

The organic matter in meteorites that have fallen to Earth mostly contain racemic mixtures of amino acids, although there is a slight enrichment in L-form amino acids giving a slight suggestion of the presence of life (Pizzarello 2004). The amino acids found in space upon the ice mantles of dust grains are thought to be racemic mixtures, with equal quantities of L and D forms (Shaw 2006, page 6). However, I do not think that the question has been fully answered yet.

Prebiotic synthesis experiments with electrical discharges to simulate lightning discharged through a strongly reducing atmosphere of gases have always produced racemic mixtures of amino acids. With no stereo-control over the carbon centre, these molecules would not form the functional molecules of cells. The question of chirality is a big problem to the hypothesized abiotic evolution of life. The so-called prebiotic chemistry that is said to have produced organic molecules that would then simply give rise to life by evolution is not so simple; a mixture of L and D configurations would inhibit the development of life by natural processes and not lead to functional cells.

The discovery of the importance of chirality in carbon-based molecules lends support to the belief that God created vast quantities of unicellular life with L-form amino acids and D-form sugars – an arbitrary decision – the importance being that the design was uniform for all forms of life. These primary unicellular forms of life gave rise to all the other forms of life continuing the production of homochiral molecules.

Meteorites do not show homochirality because they have undergone heating. Upon cooling of the meteorite, amino acids have reformed into both L and D configurations together with tar-like products. Racemization also occurs naturally over time in organic molecules. The amino acid aspartic acid undergoes racemization with a half-life of 800 years at 300 K. Thus, the age of the meteorite would be important in the question of chirality.

There is plenty of evidence of racemic organic molecules of both L-form and D-form of extraterrestrial origin. However, the lack of homochirality does not prove that these complex organic molecules were originally produced by natural physical processes. There remains the possibility that they are the transformed decay products of life.

In addition to this conclusion, a lack of homochirality indicates that these molecules could not have served as the prebiotic precursors to the evolution of life on Earth by natural processes.

Conclusion

In this chapter I have proposed that single-cell life was created in interstellar clouds on a vast scale. This life, if it exists, is intimately associated with silicate dust grains that may serve various purposes for a filamentous form of life. Silicate particles may act as solar panels to harvest light energy for metabolism or they may capture down-graded photons for photosynthesis.

The hypothesis that I presented in Chapter 10 that life has produced water applies to life in space as much as to life on Earth. According to the Life Produced Water Hypothesis the presence of abundant water is a signature of life, especially anaerobic microbial life. If water is mainly produced as a by-product of metabolism, then water is a biomarker for life. It is known that galactic space has interstellar molecular clouds containing abundant water.

It is my belief that comets are samples of interstellar dust and ice that have entered the solar system. I see comets as giant 'snow ball' conglomerates of interstellar dust grains containing organic molecules and water ice rolled together within dark nebulas by shock waves from supernovas. Comets may contain viable microbial life.

Carbonaceous chondrite meteorites contain evidences of organic life, although this organic matter has been subjected to heat during transit, so no life could have survived the journey to Earth.

It is also my belief that planetary rings are remnants of interstellar dust clouds surrounding failed stars. Variation in particle sizes is observed in the rings because they are close enough to Earth to make observation possible. Interstellar clouds are only observed by spectroscopy, so details that are not shown up by spectral analysis are not seen.

I have noted the destructive phenomena which can obliterate life in space. I believe that the small organic molecules found in interstellar clouds are not the result of prebiotic photochemistry, but of postbiotic decomposition. Small molecules would once have been the subunits of macromolecules that formed the components of cells.

The breakdown of complex organic molecules produced by living cells into simpler molecules by natural processes would be an entropic process i.e. organization moving towards disorganization. This diagnosis of the contents of interstellar molecular clouds accords with the concept of Entropic Evolution.

The overall view is that the nanocreation of life was first located in interstellar space.

Bibliography

Araki, Suguru (1991) Dynamics of Planetary Rings. *American Scientist* Jan-Feb. 1991, pages 44-59

Atlas, Ronald M., R. Bartha (1981) *Microbial Ecology: Fundamentals and Applications* Addison-Wesley Publishing Company

Canby, Thomas Y. (1993) Bacteria: Teaching Old Bugs New Tricks. *National Geographic* August, pages 36-60.

Cassé, Michel (2003) *Stellar Alchemy: The Celestial Origin of Atoms* Cambridge University Press

Crick, Francis (1981) *Life Itself: Its Origin and Nature* MacDonald & Co. London & Sydney

Davies, Paul C.W. (2008) Fitness and the Cosmic Environment. In *Fitness of the Cosmos for Life: Biochemistry and Fine-Tuning* J. Barrow et al. eds. Cambridge University Press. Pages 97-113.

De Duve, Christian (2008) How biofriendly is the universe? In *Fitness of the Cosmos for Life: Biochemistry and Fine-Tuning*. John D. Barrow, et al. eds. Cambridge University Press. Pages 169-196.

Denton, Michael J. (1998) *Nature's Destiny: How the Laws of Biology Reveal Purpose in the Universe.* The Free Press

Encyclopedia Britannica 2011 Standard Edition: Halley's Comet; Saturn: The ring system; Solar system: Formation of ring systems.

Hoyle, Fred and Chandra Wickramasinghe (1978) *Lifecloud: The Origin of Life in the Universe* J. M. Dent & Sons Ltd

Hoyle, Fred and Chandra Wickramasinghe (1993) *Our Place in the Cosmos: Life did Not Begin on Earth – It Arrived from Space and is Still Arriving* J. M. Dent Ltd. Phoenix Publications

Kandel, Robert (2003) *Water from Heaven* Colombia University Press

Klemperer, William (2008) Fine-tuning and interstellar chemistry. In *Fitness of the Cosmos for Life: Biochemistry and Fine-Tuning* J. Barrow et al. eds. Cambridge University Press. Pages 366-383.

Krieg, Noel (2005) *Bergey's Manual of Systematic Bacteriology* US Springer
(Archaea: pages 21-26).

Mader, Sylvia S. (2007) *Biology* 9th edition McGraw-Hill

Orgel, Leslie E. (2008) The Implausibility of Metabolic Cycles on the Prebiotic Earth. PLoS Biology Vol.6(1) e18 do: 10.1371/journal.pbio.0060018

Ourisson, Guy (2008) Coarse-tuning in the origin of life? In *Fitness of the Cosmos for Life: Biochemistry and Fine-Tuning* J. Barrow et al. eds. Cambridge University Press. Pages 421-439.

Piirola, V., F. Scaltriti, and G.V. Coyne (1992) Circumstellar disks deduced from sub-arcsecond polarization observations of two young stars *Nature* Vol. 359, pages 399-401.

Shaw, Andrew (2006) *Astrochemistry: From Astronomy to Astrobiology* John Wiley & Sons Ltd.

Williams, R. J. P. (1985) The Symbiosis of Metal and Protein Function *European Journal of Biochemistry* Vol. 150, pages 231-248.

Williams, R.J.P. and J.J.R. Frausto da Silva (2008) Evolution revisited by inorganic chemists. In *Fitness of the Cosmos for Life: Biochemistry and Fine-Tuning* J. Barrow et al. eds. Cambridge University Press. Pages 456-489.

www.Wikipedia.org/wiki/Panspermia

www.molres.org/astrochymist/astrochymist_ism.html

Zeilik, Michael (2002) *Astronomy: The Evolving Universe* 9th edition Cambridge University Press

CHAPTER 12

FUNDAMENTAL AND EVOLVED ELEMENTS

Introduction

Elements are types of atom. The heavy elements i.e. all elements except hydrogen, helium, lithium, beryllium and boron make up only 1 % of all atoms in the universe. A certain number of elements, maybe 24 or so, are the components of life. This chapter addresses the question; why were all the elements necessary for life conveniently found together in the right proportions to form living cells?

The current theory proclaims that the elements of life were cooked up in stars by nuclear fusion. At the risk of being extremely controversial I question the reality of some of these unobservable processes. I present the view that God created about 12 elements as the components of life with the addition of a few other elements created as catalysts. In Chapter 11 we saw that the location for this life would be interstellar clouds. Unicellular life resembling Archaea and bacteria would have been created on a vast cosmic scale. Not all carbon-based molecules in interstellar clouds would be involved in life; many would represent the debris of life – the dissociated pieces of complex molecules.

The view I am proposing in this chapter allows for the transformation of elements into different isotopes and the evolution of elements into new elements by natural processes. The Periodic Table of Elements is understood, according to this view, by dividing elements into 'fundamental elements' and evolved elements.

An alternative view could be that life was placed in a garden of elements created in the right proportions. The view presented here is that life was the supplier of elements.

One of the most surprising things is the use made by cells of metals and transition metals. I go into the details of the biological uses of metals because despite their inorganic appearance, metals have special properties that are ideally suited to specific metabolic processes. Life is totally dependent upon the properties of metals as well as upon the properties of its basic elements.

1. The elements of life

The atoms that make up living cells consist mainly of six elements: carbon, hydrogen, oxygen, nitrogen, phosphorus and sulphur. There are also other elements that play important roles in living cells; sodium, chlorine, potassium, calcium and magnesium are bulk biological elements, while the trace elements include iron, other transition metals, fluorine and iodine. Primitive organisms such as Archaea have a lesser number of elements than higher organisms, but also some elements that ceased to be used by higher organisms. Williams & Frausto da Silva (2008) list some 21 elements in total necessary to all forms of life.

Metal ions

Many activities performed by cells are dependent on ions. Ions are electrically charged atoms that float freely in the cytoplasm of cells or in the fluid surrounding cells. The metal ions of sodium, potassium, magnesium and calcium are smaller than organic compounds; they thus diffuse quickly in aqueous solutions and often serve in the transmission of information. The uses of various elements in prokaryotic bacteria are further developed and extended in multicellular organisms. These are the uses of the principle metal ions:

Sodium (Na^+) and **potassium** (K^+) ions are used to create electrical gradients across membranes. This forms a basis for the conducting of nerve impulses in neurons and muscle fibres. **Calcium** (Ca^{2+}) ions are used in chemotactic communication between single-celled organisms and in signalling between the cells of multicellular organisms. Calcium ions play a role in muscle contraction, hormone release and the transmission of nerve impulses across a synapse among other things. **Chlorine** (Cl^-) nonmetal ions play an inhibitory role in cell systems.

The transfer of sodium ions in and out of cells is also important in osmotic control in animal cells.

Therefore, metal ions (and a few nonmetals such as chlorine) play important roles in creating electrolytic signals within and between

62

cells and in osmotic balance. They are also the principal components of enzyme catalysts.

The macromolecules of cells are synthesized mainly from the following small molecules as raw materials: methane (CH_4), water (H_2O), carbon monoxide (CO), carbon dioxide (CO_2), hydrogen sulphide (H_2S), ammonia (NH_3) and hydrogen cyanide (HCN). The biological synthesis is accomplished by the reduction of some of these molecules and the reaction of all of these small molecules using catalysts that have metal ions in their composition (Williams & Frausto da Silva 2008, page 461).

Magnesium (Mg) is used in various important enzymatic activities. Chlorophyll is composed of a long side chain of hydrocarbons and four carbon rings around a central magnesium atom. Photosynthesis depends on the light-absorbing properties of magnesium which are several thousand times higher than iron. It is probable that the light-absorbing capacity is due to the electronic structure of magnesium. The formula of the chlorophyll porphyrin is $C_{55}H_{72}O_5N_4Mg$.

Magnesium is an important component of ATP, the energy-storage molecule. ATP should be written Mg-ATP. Three hundred enzymes in human cells use magnesium ions for their catalytic action. These include the enzymes that synthesize DNA and RNA. The CDK enzyme has an Mg^{2+} ion used in phosphorylation to activate the protein kinase enzyme.

Transition metals
Various transition metals are essential to the action of enzymes, the principal one being iron. The transition metals have complex electron shells that give them electrical conducting properties enabling them to trap and channel electrical energy. This gives them a vital role in cell metabolism, as well as in modern technology.

"It is only by utilizing the conducting properties of the transitional metals that the cell is able to channel the electron flow through discrete energy steps and utilize each energy drop to perform useful chemical work. No organic compounds can substitute for the transitional metals in this regard." (Denton 1998, page 199-200).

Iron (Fe) is attached to cyclic carbon compounds called heme and used in oxidation-reduction (redox) processes. The attachment of the

Fe^{2+} ion to heme groups is by chelation. Chelation of iron, for example by the adding of hydroxyl and carbonyl groups, is used by some types of bacteria to solubilise iron and take it into the cell for various purposes. However, the best-known iron chelates are cytochrome c oxidase and haemoglobin. Cytochrome c oxidase is an enzyme found in mitochondria and haemoglobin is the oxygen-carrier of blood, both are involved in oxidative metabolism.

Oxidative metabolism or aerobic respiration takes place in the mitochondria of cells. It involves the production of ATP with energy released in the citric acid cycle or Krebs cycle. Cytochrome metalloproteins have heme groups each with a central atom of iron that perform an electron transport chain. The iron atoms become reduced when they accept electrons, and oxidized when they give up electrons. Energy is released at each step of the electron transport chain and the spent electrons are passed from the last carrier to oxygen that is transformed into water (Mader 2007, page 138).

Haemoglobin consists of four heme groups made from the protein globin each with a chelated iron atom. The properties of the iron atom allow it to associate with molecular oxygen in the lungs, convey oxygen (which is a highly reactive molecule) safely in the blood to the tissues where it dissociates from the oxygen, releasing the oxygen to the mitochondria of the cells.

Vertebrates use this system to convey oxygen to their cells. Some invertebrate arthropods and mollusks have haemolymph containing haemerythrin, an unbound form of iron instead of haemoglobin. Other marine invertebrates use haemocyanin containing copper atoms for the binding of oxygen.

Nickel (Ni) is the principal component of a yellow tetrapyrrole coenzyme F_{430} that is unique to methanogenic Archaea. It catalyzes the last step of the pathway that produces methane. Nickel has numerous other biological roles in microorganisms and plants. The aerobic bacterium *Aquifex* uses hydrogenase enzymes containing nickel and iron-sulphur clusters to oxidize hydrogen to water. Another bacterium uses carbon monoxide dehydrogenase enzymes containing iron-nickel-sulphur clusters.

Nickel is found in the urease enzymes of plants, but it is absent in animals. Thus, nickel had widespread use in the metabolism of primitive organisms, but has little use in later multicellular organisms.

Cobalt (Co) is used by primitive organisms in carbon transfer. Cobalt is a central component of the vitamin cobalamin, or vitamin B_{12} in animals. There are a few cobalt proteins derived from vitamin B_{12} but the use of cobalt is rare in higher organisms.

The cobalt of cobalamin and the nickel of F_{430} factor are chelated. The chelation of metal ions is an irreversible process that prevents these atoms from being replaced by other free metal ions that would thereby disrupt the system.

Vanadium (V) and **tungsten** (W) played important roles as catalysts in primitive anaerobic bacteria. Vanadium is a component of the nitrogenase enzyme of anaerobic bacteria. Tungsten takes part in the redox processes performed by extremophile Archaea. They were both displaced by molybdenum in aerobic bacteria.

Molybdenum (Mo) is found in nitrogenase, an enzyme used in the reduction of nitrogen to ammonia in aerobic bacteria. It is also the active ingredient of nitrate reductase which catalyzes the reduction of nitrate to nitrite. Only bacteria fix atmospheric nitrogen, allowing it to be incorporated into organic compounds such as amino acids. Nitrogenase also contains iron.

Manganese (Mn) ions are used in the detoxification of superoxide free radicals. All aerobic organisms possess enzymes whose job it is to eliminate reactive oxygen radicals. Manganese may serve well this purpose due to its many oxidation states of +2, +3, +4, +6 and +7. A metalloenzyme containing four manganese atoms is also important to aerobic photosynthesis using photosystem II.

Copper (Cu) was not widely available under anaerobic conditions because it forms insoluble sulphides in the absence of oxygen and water. The biochemistry of primitive anaerobic microorganisms included only about 15 elements. They used nickel and cobalt rather than copper and zinc which became important later on.

In aerobic conditions reactive oxygen radicals cause chemical changes in proteins that make them non-functional. Cells must be protected from the destructive effects of oxygen by enzymes and free-radical (OH) scavengers. Superoxide dismutase is a protector enzyme that occurs in all aerobic organisms. In prokaryotes it uses iron and manganese or nickel, but in eukaryotes the enzyme relies on the properties of copper and zinc. Multicellular organisms also use copper and zinc for protection against invasive organisms. Protection from disease in animals involves the heme iron of thyroid hormones, oxidized **selenium** (Se), and selenium and **iodine** (I) combined.

Copper is found in a variety of enzymes. Besides iron, the cytochrome c oxidase enzyme of oxidative metabolism has a central copper ion. Zinc and magnesium also play a structural role in cytochrome c oxidase.

Cross-linking in animal tissues is accomplished by oxidases that are copper enzymes found in extracellular fluids; they guide the production of collagen. Cell to cell communication in multicellular organisms is accomplished by chemical messengers that are organic compounds called hormones. Hormones act outside cells, sending messages across the cell membrane by calcium pulses. Hormones such as adrenaline and amidated peptides are synthesized using copper enzymes.

Most mollusks and crustaceans have the copper-containing pigment haemocyanin rather than the iron-containing haemoglobin for oxygen transport round the body. Their blood or haemolymph is blue.

Zinc (Zn) has many important biological functions in multicellular organisms. Zinc forms part of many DNA transcription factors, one is called the zinc finger. Zinc as a cofactor is an activator of certain enzymes. It plays a role in disrupting the laying down of tissues containing collagen so as to allow for growth of the body, and the migration and reassembling of animal cells.

Many cells in the body secrete zinc ions (Zn^{2+}) as signals; these act as inorganic hormones. Zinc ions are also considered to be neurotransmitters in the brain. The polymerization of nucleic acids to form DNA and RNA is performed by enzymes containing zinc ions. Olfactory receptors that allow the sense of smell have zinc binding sites. Zinc is found in red blood cells as part of the enzyme carbonic anhydrase.

In nut-bearing and fruit trees zinc deficiency causes various diseases such as mottle leaf. Yeast uses zinc in alcoholic fermentation. Zinc functions in the hemosycotypsin of snail's blood to transport oxygen in the same way that iron in haemoglobin transports oxygen.

Nonmetals and structural elements

There are also elements whose role is in the strengthening of outer structures. Molecules secreted by cells to form protective structures principally contain the nonmetal silicon and the metal calcium.

Silicon (Si) is the main constituent of the silica ($Si(OH)_4$) cages of the protist Radiolaria; as well as the tiny shells of diatoms –single cell, motile algae and silicoflagellates. These same silicates are the raw material of the rock called chert or flint. Grasses of the plant kingdom also incorporate silica to strengthen and protect their leaves.

Calcium (Ca) is deposited by a wide variety of organisms as a matrix, shell or skeleton. Cyanobacteria deposit calcium carbonate ($CaCO_3$) within its bacterial mat. Protist Foraminifera have a test (small shell) composed of foreign matter cemented together with calcium carbonate. Golden, green and red algae and corals build calcium carbonate reefs. Some sponges have calcareous skeletons and silica spines, others have calcareous spines. Mollusk bivalves and gastropods (snails) have calcium carbonate shells. The brachiopods (Lamp shells) have calcium phosphate ($Ca_3(PO_4)_2$) shells.

The bone of vertebrates is an extracellular mineralized matrix strengthened by crystals of apatite ($Ca_5(PO_4)_3OH$) but with an important carbonate content and a small amount of calcium phosphate. Thus, **phosphorus** as well as calcium is an element contributing to the skeletal structure of multicellular organisms.

The enamel of teeth contains **fluorine** (F) as fluorapatite ($Ca_5(PO_4)_3F$).

Fluorine is a halogen, as is chlorine. Gastric juice in the digestive tract of animals contains **chlorine** as hydrochloric acid (HCl).

In conclusion, I have counted 24 elements as the total number used by all forms of life. Three of these elements are only used by bacteria;

they include molybdenum (Mo), vanadium (V) and tungsten (W). Five are mainly used by multicellular organisms, they include copper (Cu), zinc (Zn), selenium (Se), iodine (I) and fluorine (F). Silicon (Si) is used exclusively in the formation of outer structures. The list of 24 includes eight nonmetals, four metals, nine transition metals and three halogens. Williams & Frausto da Silva (1985, 2001) list some possible trace elements for bacteria, plants and animals in addition to the ones already mentioned. This adds chromium (Cr), arsenic (As), bromine (Br), strontium (Sr), cadmium (Cd), tin (Sn) and barium (Ba) to the list I have made.

2. Periodic Table of Elements

The Periodic Table of Elements places the elements in rows; each element to the right has one more positively charged proton in the nucleus and one more negatively charged electron orbiting the nucleus than the previous element. The elements are also placed in columns; the elements of a column have the same number of electrons missing from the outer electron orbital shell. This number of missing electrons determines the number of bonds the element can make with other elements, and is known as valence or valency.

The valency is important to the chemical properties of elements. The columns of elements are classed broadly into metals on the left, nonmetals on the right and the transition metals in between.

Metals undergo three-dimensional bonding of similar atoms that gives multicentre orbitals. The metal atoms form a lattice of positively charged nuclei surrounded by electrons moving around the lattice. This allows electron mobility over the whole crystal which is the basis to electrical conductivity in metals. The transition metals are small, hard atoms that are less volatile compared to the metals. Iron, nickel and cobalt have magnetic properties. Platinum (Pt), silver (Ag), gold (Au) and mercury (Hg) are the least reactive of transition metals. The halogens are reactive nonmetals including fluorine (F), chlorine (Cl), bromine (Br) and iodine (I).

The far-right column includes the inert gases also known as the noble gases helium (He), neon (Ne), argon (Ar), krypton (Kr), xenon (Xe) and radon (Rn). They have a complete outer electron shell which means that they do not form bonds and so are unable to undergo chemical interactions and form molecules. They exist as isolated atoms.

Hydrogen stands at the top of the chart, alone, with one proton and one electron. It is different from every other element in that it has no neutrons (although its isotopes deuterium and tritium have one and two neutrons respectively).

The rows of elements are numbered 1 - 7. This indicates the number of electron shells around the nucleus. There is a maximum of seven electron shells representing the maximum distance that electrons can orbit from the nucleus. The outer (and only) electron shell of row 1 with H and He is completed with two electrons. The outer electron shell of rows 2 and 3 are completed with eight electrons. The outer electron shells of rows 4 and 5 are completed with 18 electrons. Finally, the outer electron shells of rows 6 and 7 are completed with 32 electrons.

Uranium (U) located in row 7 with 92 protons, 92 electrons and higher numbers of neutrons is the largest of naturally occurring atoms. Elements are only stable up to bismuth (Bi) which has 83 protons. All the elements above bismuth including all those in row 7 are unstable and transform themselves into lead (Pb) with 82 protons. There are man-made elements above uranium including plutonium (Pu) that are extremely unstable.

12.1 PERIODIC TABLE SHOWING THE ELEMENTS MENTIONED IN THE TEXT AND ESPECIALLY THOSE OF WHICH LIFE IS COMPOSED

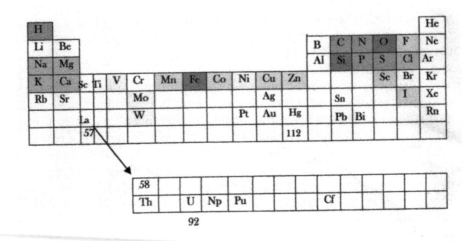

- Catalysts in star formation
- Main elements of life
- Other bulk biological elements
- Trace elements
- Elements used only by bacteria

A new understanding of the Periodic Table

Notice that the 20 or so elements that are the constituents of living cells are all found at the top of the Periodic Table. They come from different groups of elements and so have almost the full range of chemical properties. The metals and transition metals are among the elements important for life processes. The 24 most abundant elements are all associated with living cells with the exception of three inert gases. The inert gases helium, neon and argon are abundant but being inert, they play no role in organic chemistry.

After hydrogen and helium, the cosmic abundance of elements ranks oxygen at three and carbon at four. Iron is abundant in the universe; sodium, potassium, calcium, magnesium and copper are also abundant.

The extraordinary coincidence between the elements involved in life and their greater abundance than other elements in the universe caused me to start to work on a new way of understanding the Periodic Table of Elements. It occurred to me that there could be 'fundamental elements' and evolved elements. The fundamental elements would have been created as the components of various life forms, their various properties serving specific purposes.

The evolved elements would come about by the natural processes of nucleosynthesis. Evolved elements are found lower down in the Periodic Table and they are often characterized by instability. The evolved elements are also subject to random processes, the main one being radioactive decay.

The evolved elements are of four main types; the radioactive elements lower in the chart; the radioactive isotopes of all elements; elements produced by the splitting of the nucleus in nuclear fission; and the inert gases which once produced, accumulate due to extreme stability in their atomic configuration.

I propose that at least 12 elements were created as the components of life: they include the bulk organic molecule elements; the cell function elements; and the metal/transition metal component of enzymes. In addition there are elements secreted by cells to provide support and protective structures, and involved in digestion.

Hydrogen already existed. It was the first element formed in the universe, in a sense it *is* the universe. Later hydrogen became an important element for life.

Apart from the elements created for life, three others were created as catalysts – these three elements are iron, oxygen and silicon. Iron is now found in planetary cores, and silicon and oxygen as silicates form the rocky mantles and crustal surfaces of planets. (The catalytic role of iron and silicate interstellar dust will be discussed in Chapter 13, section 1).

Aluminium is not a component of life; it is found as aluminosilicate minerals and constitutes 8 % of the Earth's crust. Aluminium is near the top of the Periodic Table and quite abundant, but I have not included it as a fundamental element; it may have evolved from magnesium or silicon.

The **Fundamental Elements – Components of Life Hypothesis** is as follows:

Fundamental elements (with the exception of hydrogen, oxygen, silicon and iron) were created as the constituents of living cells inhabiting interstellar dust clouds as unicellular forms of life. The properties of the fundamental elements were designed to sustain life. The fundamental elements found at the top of the Periodic Table of Elements have evolved into the elements found lower down in the chart, as well as into unstable isotopes of each element and inert gases by natural processes.

Therefore, according to the Fundamental Elements – Components of Life Hypothesis, the Periodic Table of Elements consists of fundamental created elements and elements that evolved from the fundamental elements by natural processes.

Radioactivity

Radioactivity was discovered by the French physicist Henri Becquerel (1852-1908) in 1896 while working on uranium. Marie Curie (1867-1934) Polish-born French physicist named the phenomenon. She and her husband Pierre Curie (1859-1906), French physical chemist, discovered radium and polonium. Marie Curie was the first woman to win a Nobel Prize in 1903 shared with her husband and Henri Becquerel.

Atoms are only stable up to element 83, bismuth (Bi). The elements beyond this break down by a series of often short-lived radioactive elements into element 82, the heavy metal lead (Pb). Atomic mass varies according to the number of neutrons in the nucleus. Different isotopes have different atomic masses and numbers of neutrons. The atomic mass is written as a number before the symbol e.g. ^{238}U.

Uranium is the heaviest naturally-occurring element with an atomic number of 92. Neptunium (93) and plutonium (94) also occur naturally, but only in trace amounts. The plutonium of atomic bombs is man-made. There are another 18 man-made elements above uranium with the highest atomic number being 112. The radioactive decay of the elements above uranium is variable. The half-life may be as long as tens of millions of years. Californium (Cf) at 98 has a half-life of 800 years. Some man-made elements only exist for mere fractions of a second and any elements made beyond atomic number 112 would simply fly apart.

^{238}Uranium undergoes successive radioactive decays in 14 steps until it is transformed into stable ^{206}Lead. Uranium with atomic mass 235 decays into ^{207}Lead. Thorium with atomic mass 232 becomes ^{208}Lead. These decays include the inert gas radon (Rn) in the series. All lead except ^{204}Pb is the end product of a complex radioactive decay.

Radioactive decay is a nuclear process whereby an unstable nucleus emits alpha particles, beta particles and gamma rays. Alpha particles are two protons with two neutrons. Beta decay is the expelling of a high-speed electron, and gamma rays are electromagnetic waves. Radioactive decay causes the nucleus to lose mass and thus change into an element with a lower mass and proton number. The kinetic energy carried by alpha and beta decay particles heats up material such as rock surrounding radioactive elements.

It has been possible to produce at least one radioactive isotope for every known element in the Periodic Table by bombardment of the nucleus.

Isotopes

We have already mentioned the word isotope. Isotopes of an element are versions of the element with different numbers of neutrons in the nucleus. Different atomic masses for uranium and lead have been mentioned. The atomic mass of an element is the sum of the number of protons and neutrons in the nucleus. Thus, different isotopes of an element have different atomic masses due to different numbers of neutrons, but the same atomic number which is the number of protons in the nucleus. The gain or loss of protons in a nucleus causes the element to become another element.

The elements of atomic number 1 to 83 (hydrogen to bismuth) have on average three isotopes, although tin (Sn) has ten isotopes.

Isotopes with greater numbers of neutrons than protons have unstable nuclei. Carbon with atomic number 6 has three isotopes. ^{12}C is a stable isotope with six protons and six neutrons. ^{13}C and ^{14}C have additional neutrons and are subject to decay.

^{14}C is formed in the atmosphere by the bombardment of the Earth by neutrons. The collision of neutrons with ^{14}N (the common isotope of nitrogen in N_2) transforms ^{14}N into the unstable, radioactive isotope of carbon ^{14}C. The constant occurrence of this process maintains a nearly constant concentration of carbon dioxide containing ^{14}C in the atmosphere. Carbon dioxide containing both ^{12}C and ^{14}C is fixed by plants practicing photosynthesis to produce a ^{12}C / ^{14}C ratio in living things. Over time ^{14}C decays back into the more stable ^{14}N (Shaw 2006, pages 166-168). Thus, over time the carbon-14 content in materials is reduced.

Radiometric dating

^{14}C has a half-life of 5717 years. Half-life is the length of time for half the radioactive material to disintegrate into another element. If the amount of the radioactive isotope contained in a material is known and the half-life is known, the amount remaining of the radioactive isotope in a sample allows the age of the sample to be calculated. Carbon dating is used for time scales in the recent past, for example, in the dating of historical artefacts.

^{238}Uranium has a half-life of 4.5 thousand million years. Uranium/lead dating is used for samples with a geological time scale such as the dating of meteorites or geological rock formations. Rubidium/strontium dating involving ^{87}Rb and ^{87}Sr and a half-life of 48.8 thousand million years is not useful for most samples. Potassium/argon dating involving ^{40}K and ^{40}Ar and a half-life of 1.25 thousand million years has more application. Thorium/lead dating involves ^{232}Th and ^{208}Pb and a half-life of 14 thousand million years. The age of stars is measured by the abundance of thorium.

The radioactive isotope of Aluminium ^{27}Al decays to magnesium ^{26}Mg with a half-life of 730 000 years. Comparison of the $^{26}Mg/^{27}Mg$ ratio allows the age of chondrules in carbonaceous chondrite meteorites to be measured. It has led to the conclusion that the meteorites are about 1.2 million years older than the chondrules they contain (Shaw 2006, page 168). Most magnesium is ^{24}Mg.

3. Nucleosynthesis

The theory of stellar nucleosynthesis was pioneered by Fred Hoyle (1915-2001) and William Fowler[2]. In the mid-1950s they presented a cosmological framework in which the stars act as nuclear furnaces forging new elements out of hydrogen and helium. Prior to this it had been thought that the different elements had been made in the Big Bang.

Hydrogen and helium are known as 'light elements'. The elements lower in the Periodic Table than boron are 'heavy elements'. The heavy elements are also confusingly referred to as 'metals', although they are not all metals. The reason for this is that the measurement of iron content compared to hydrogen (Fe/H) is taken to indicate the proportion of all heavy elements in a star. The proportion of heavy elements in a star's atmosphere is the metallicity (Z) of the star. Iron is very abundant in the universe; it has the highest binding energy of all nuclei and is able to resist the highest temperatures.

The aims of nuclear astrophysics are to understand how nuclear reactions generate heat and light in stars; and how simple elements can be transmuted into more complex ones by processes of nucleosynthesis.

Spectroscopy
The science of spectroscopy was first discovered by the Italian Jesuit Priest and astrophysicist Pietro Angelo Secchi (1818-1878). In 1849 Secchi became the director of the Vatican Observatory where he made the first survey of the spectra of stars and suggested that stars could be classified according to their spectral type. The theory behind spectrum analysis was established by Gustav Robert Kirchhoff (1824-1887), a German physicist and Robert Bunsen (1811-1899), a German chemist in work commencing in 1859.

Spectroscopy allows the chemical abundances to be assessed for the surface layers of stars. This has been used for stars within our own galaxy, and is complemented by abundance measurements for stars in other galaxies even at high red shifts. It has been found that older stars have lower metallicity than younger stars. The older stars (which

---

[2] Burbidge, E. M., G. R. Burbidge, W. A. Fowler and F. Hoyle (1957) Synthesis of the Elements in Stars. *Review of Modern Physics* Vol. 29, pages 547-650.

formed earlier) are found in the halo era of galaxies –the spherical-shaped galaxy when it first formed. These stars are classified as Population II stars. They contain only between 0.01 and 0.02 % of their mass in heavy elements (Zeilik 2002, page 349). The younger stars aged between a few tens of millions of years old to ten thousand million years are found in the disk era of galaxies (galaxies have often evolved into disk shapes). They are classified as Population I stars. Population I stars contain 1 to 2 % of their mass in heavy elements.

The observation that stars of different ages do not have the same proportion of heavy elements indicates that the Big Bang could not be the origin of all elements. The gradual enrichment in metals or heavy elements in stars over time has suggested a process of chemical evolution. But as Cassé (2003, page 78) rightly points out, "It is one thing to observe and record the state of atomic matter in the Universe, and quite another to explain it."

Hydrogen fusion inside stars

Hydrogen burning inside main-sequence stars produces helium. Main-sequence is a term given to stars that have nuclear reactions going on inside. This happens during most of their life time, while protostars are baby stars that have not started up their nuclear furnace and stellar corpses are stars where nuclear reactions have ceased (star evolution will be explained in Chapter 13). The conversion of hydrogen to helium produces heat and starlight.

The main fusion reaction converting hydrogen into helium is the proton-proton chain (*PP chain*). When the core of a star such as the Sun reaches 8 million K and high density, hydrogen nuclei can collide with enough energy for the protons to stick together. A heavy hydrogen nucleus (^2H) forms. One proton converts into a neutron by the positive charge breaking away as a positron. Meanwhile a neutral, supposedly massless particle called a neutrino (v) also breaks away and exits the star carrying away some energy. The positron quickly collides with an electron to form two gamma rays (gamma rays are photons). The heavy hydrogen (^2H) collides with another proton and forms light helium (^3He) and a gamma ray. Then light helium nuclei collide whereby two light helium nuclei fuse to form ordinary helium (^4He) plus two protons and another gamma ray.

Each completed *PP chain* unlocks 4.2×10^{-12} J. The Sun converts 4.5×10^6 metric tons of matter into energy every second. The gamma ray photons which carry most of the star's energy cannot exit the star directly. The gamma rays bounce between the star's atoms being absorbed and reemitted. The original gamma rays are degraded into many lower-energy photons that finally leave a star such as the Sun as visible light after about 200 thousand years of bouncing around inside.

A star such as the Sun will burn the hydrogen supplies in its core which represents about 10 % of its total hydrogen converting 0.7 % of the mass into radiant energy. At the end of its life, the core will have been converted to helium. The Sun's size and hydrogen supply will allow nuclear fusion reactions to go on for 10 thousand million years, and it is 5 thousand million years old. Our star will not lose its brightness for a while yet.

Thus, the nuclear fusion reaction performed by bright stars involves converting protons into neutrons, emitting light energy as gamma rays that become visible light and ultraviolet light, and producing the inert gas helium. Helium is also thought to have been produced by the initial thermonuclear event of the Big Bang giving a universal helium abundance of about 23 %.

Supernova synthesis
It is known that the elements in the Periodic Table beyond iron are not produced by nuclear fusion. This is because the positive charge of the nuclei in these heavier elements could not be overcome by temperature since temperatures that high would cause the nucleus itself to disintegrate. Consequently, the strong electrical repulsion of the nuclei of heavy elements rules out nuclear fusion as a mode of formation.

The elements beyond iron are thought to be synthesized by successive neutron captures followed by the transformation of neutrons into protons by beta decay (β decay). Neutrons easily penetrate a nucleus because they carry no charge. Excess neutrons decay into protons by beta decay whereby a neutron converts into a proton, electron and antineutrino. Neutrons have a lifetime of about 10 minutes outside the nucleus.

High energy neutrons that bombard and enter the nuclei of iron and other metals are produced mainly by supernova explosions. Dying stars collapse becoming neutron stars exploding in the process. When the outer layers are blown off the event is known as a nova, and when the whole star explodes including the core it is known as a supernova. The ejection of neutrons into the interstellar clouds of surrounding space allows neutron capture events that produce elements with higher atomic numbers.

There are two types of neutron capture; rapid neutron capture known as the *r process* and slow neutron capture known as the *s process*. The *r process* occurs during supernova explosions and results in most of the elements heavier than iron. In the *r process* neutrons are captured at a rate faster than the beta decay rate, so neutron-rich nuclei are formed. In the *s process* neutrons are captured at a rate slower than beta decay, so proton-rich nuclei are made (Zeilik 2002, page 373).

Explosive nucleosynthesis produces the elements gold, silver and platinum. The heaviest element it produces is uranium. Shaw (2006, page 97) also names copper, zinc, palladium, bromine and iodine, as well as the lanthanides and actinides (radioactive elements at the bottom of the Periodic Table) as elements formed during supernova explosions.

Slow neutron capture by iron, the *s process*, occurs on the surface of some red giant stars. It is thought to produce some of the elements between iron and lead. In the slow process neutron captures are separated by neutron $\rightarrow$ proton conversions.

The elements beyond lead are readily affected by radioactive decay which is a nuclear fission process.

Cosmic ray fission
The elements between helium and carbon are rare in the universe. They are lithium (Li), beryllium (Be) and boron (B) with atomic numbers 3, 4 and 5. It is now understood that they are produced by nuclear fission known as spallation.

Spallation events occur at high altitude in the Earth's atmosphere and in the interstellar medium of the galaxy, and involve cosmic ray collisions. Cosmic rays are highly accelerated particles; 85 % of galactic cosmic rays are protons –the nuclei of hydrogen atoms, 12 % are alpha particles –the nuclei of helium atoms, and the remaining 3 % are the nuclei of heavier atoms. When cosmic rays consisting of carbon or oxygen nuclei collide with atoms in the upper atmosphere or in interstellar clouds, they fragment producing cosmic ray nuclei of lithium, beryllium and boron. This makes the abundance of Li, Be and B in cosmic ray populations greater than expected (*Encyclopedia Britannica 2011 Standard Edition* : Cosmic ray). So, the nuclear fission production of Li, Be and B occurs by the shattering of C and O nuclei under the effect of violent collisions.

4. Do stars burn their own ashes?

The currently accepted theory is that massive stars produce the elements from carbon to iron by nuclear fusion. These nuclear fusion reactions are said to occur when hydrogen fusion stops and the star is dying. They are of very short duration –often a thousand years or less.

Temperature is the key to these fusion reactions. Atomic nuclei carry positive electrical charges that strongly repel other nuclei and prevent them from approaching each other. The larger the number of protons in the nucleus, the stronger the positive charge and the greater the repulsive force. It is only at temperatures of millions of degrees that the positively charged nuclei could gain sufficient kinetic energy to overcome their mutual electrical repulsion allowing them to combine under the attraction of the short-range nuclear force.

Helium, carbon, neon, oxygen and silicon burning
It is believed that fusion reactions form the elements at the top of the Periodic Table. The fusion reactions that would form heavier elements start with helium burning where three helium nuclei fuse to form one carbon. Helium burning is said to produce both carbon and oxygen, and occur in small stars as low as one solar mass. Helium burning is followed by carbon burning to produce oxygen, neon, sodium and magnesium. Carbon burning is followed by neon burning to produce oxygen and magnesium. After this comes oxygen burning in stars of ten solar masses to produce magnesium, aluminium, silicon, phosphorus and sulphur. Finally, silicon burning that uses as fuel the elements between magnesium and sulphur occurs to produce the

elements near to iron in stars of twenty solar masses (Zeilik 2002, page 353).

Carbon fusion that produces magnesium, sodium and neon is represented as follows:

$$^{12}C + {}^{12}C \rightarrow {}^{24}Mg + \gamma,$$
$$^{23}Mg + n,$$
$$^{23}Na + p,$$
$$^{20}Ne + \alpha.$$

γ = Gamma photon, n = neutron, p = proton, and α = helium nucleus (Cassé 2003, page 99).

Oxygen fusion that produces sulphur, phosphorus and silicon is represented as follows:

$$^{16}O + {}^{16}O \rightarrow {}^{32}S + \gamma,$$
$$^{31}P + p,$$
$$^{31}S + n,$$
$$^{28}Si + \alpha.$$

Silicon fusion cannot occur by the fusion of two silicon nuclei because the electrical charge is too high. A different method is proposed involving additions of helium nuclei.

The theory is that the interior of a massive star becomes layered like an onion due to decreasing temperature zones from the core outwards. An iron core is surrounded by successive layers of silicon, neon, carbon and oxygen, helium and hydrogen. (Cassé 2003, page 100).

In a 25 solar mass star helium burning is calculated to take 500 000 years; carbon burning 600 years; neon burning one year; oxygen burning 6 months; and silicon burning one day. Then supernova explosion occurs (Zeilik 2002, page 373).

Thus, it is claimed that stars burn their own ashes. The duration of combustion phases depends on the size of the star, and only stars of 8 solar masses or more perform all of the fusion cycles.

The successive combustion of hydrogen, helium, carbon, neon, oxygen and silicon require ever higher temperatures to overcome the electrical repulsion between reacting nuclei. The ashes from one cycle serve as the fuel for the next cycle. Each cycle produces less energy than the previous one, and lasts for a shorter time (Cassé 2003, page 96-97).

According to Cassé (2003, page 101), oxygen, neon, and magnesium are produced inside stars, while sulphur, argon, calcium and iron are produced in the supernova explosion that ends the series of processes. It is generally thought that the elements carbon to calcium are synthesized inside high-mass stars, while iron and its kin are generated in the final explosion.

Temperatures of stars

In this section I am going to look at the temperatures of stars in order to compare observation with theory.

Only the surface temperatures of stars can be measured. The core temperatures given for stars are hypothetical; they depend upon the theoretical composition of the core and the processes supposed to be taking place there. Neither the temperature nor the composition of the interior of stars is, at present, open to observation.

The surface temperatures are known and are hottest for main sequence stars. Class O stars are some of the hottest with surface temperatures of 30 000 K and a bluish-white colour. The photons that these stars emit are so energetic that they ionize the surrounding hydrogen. Class M stars are the coolest at 3500 K or less and a reddish colour. Cool, low luminosity main sequence M stars are the commonest type in our galaxy.

12.2 SURFACE TEMPERATURES FOR STARS

Spectral Class of star	Surface temperature	Examples of stars
O	30 000 K	Naos
B	30 000 – 11 000	Rigel, Spica
A	11 000 – 7500	Sirius, Vega
F	7500 – 6000	Canopus, Procyon
G	6000 – 5000	Sun, Capella
K	5000 – 3500	Arcturus, Aldebaran
M	3500 and less	Betelgeuse, Antares

(Modified from Zeilik 2002, page 292)

As already mentioned, the end of hydrogen fusion and the conversion of the core of a star to almost pure helium, mark the end of the star's main sequence life time. Stars then enter a dying phase as red giants or white dwarfs.

Red giants are generally hundreds of times bigger than the Sun. They have very high luminosities due to their large surface area, but low surface temperatures of between 2000 and 6000 K. White dwarfs have very low luminosities compared to their surface temperature due to their very small size. Cool white dwarfs have surface temperatures of 4000 K or lower, while hot white dwarfs have surface temperatures of over 25 000 K. Some have temperatures of 70 000 K and emit X-rays. White dwarfs cool down over time.

It is thought that the core of a red giant will evolve into a white dwarf in a stage called the planetary nebula (this name has historical origins and is not connected with the formation of planets). Zeilik (2002, page 352) writes of the central stars of planetary nebulas, "Some central stars are extremely hot and luminous. One measured by the HST suggests a surface temperature of 200 000 K – the hottest star known to date!"

When a red giant has blown off its outer layers (in a nova), it leaves behind a hot, dense core which becomes a white dwarf. The high recorded temperature of this star may be due to the fact that it represents the core of a red dwarf, while the surface temperature would have been much cooler.

The Sun has a surface temperature of 6000 K. Parts of the Sun's corona have very high temperatures of 2 million K causing the ionization of elements such as iron, nickel, neon and calcium. The estimated temperature of the Sun's core is 8 million K, with almost 16 million K at the very centre.

Hypothetical core temperatures in red giants
The nucleosynthesis of the elements between carbon and iron is said to occur inside red giant stars which are large size, high luminosity but low temperature stars.

As has already been stated, the electrical repulsion between nuclei inhibits fusion reactions from taking place. Cassé (2003, page 65) points out that,

"The probability that two nuclei join together decreases exponentially with the product of their electrical charges. For example, the fusion of two carbon nuclei involves a product 36 times greater than for two hydrogen nuclei, and this number occurs as the argument of an exponential. In other words, the probability is a factor of 1/exp 36 lower."

These strong inhibitions must be overcome by increasing the velocity of nuclei through increasing the temperature. The source of heat is hypothesized as being produced by gravitational contraction. The core of the red giant is said to collapse and contract, and this raises the temperature of the core.

This is a list of the necessary, but hypothetical temperatures at which the fusion cycles take place (cited from Cassé 2003, pages 70, 99-100):

- Helium fusion: 100 million K (10^8 K).
- Carbon fusion: 600 million K (page 70); slightly below a billion K (page 99).
- Neon fusion: 1000 million K (page 99).
- Oxygen fusion: 1000 million K (page 70); 2000 million K (page 100).
- Silicon fusion: 2000 million K (page 100); 3000-4000 million K (page 70).

Hydrogen fusion involving the *P-P chain* that converts hydrogen into helium occurs at 8 million K. This is the estimated temperature of the core of the Sun. Notice that these hypothetical temperatures are many orders of magnitude greater than any observed temperatures associated with stars.

The other mode of nucleosynthesis already mentioned is associated with supernova explosion. Supernovas produce shock waves which plough through the interstellar medium at speeds of hundreds of kilometres per second. They compress and heat the interstellar gas to a few million K. This hot gas is observed in X-ray images of interstellar clouds. Elements such as radioactive nickel-56 are formed at the very high temperatures produced in supernova explosions.

Fusion reactions leading to the nucleosynthesis of the elements at the top of the Periodic Table are said to occur inside red giants, yet nothing of these processes shows at the surface of the star. The massive amounts of heat are said to be produced by gravitation as the core contracts in the dying star, but the star itself is not seen to contract (Zeilik 2002, page 340). The claimed contraction of the core in red giants seems to contradict the model that represents red giants as expanding massively in size when they reach this stage.
It is also surprising that such phenomenally high temperatures could exist within a star while having no effect whatsoever on the star's luminosity or surface temperature. In main sequence stars, the thermonuclear reactions going on inside the star clearly affect the surface temperature, making them observably hotter than red giants at the surface. Cassé (2003, page 99) writes that as the core continues to heat up,

"The characteristic time-scale of nuclear evolution in the core becomes shorter than the time required by the visible envelope to readjust. It follows that the internal evolution does not result in any changes in colour or luminosity."

Again, on page 133 Cassé states with reference to helium burning,

"The nuclear reactions race ahead, freeing enormous quantities of energy in a very brief space of time. And yet this profusion of energy leaves almost no trace at the surface. Indeed, the energy produced filters very slowly through towards the distant photosphere of the red giant, and by the time it finally arrives, it has been diluted and softened."

Let it be noted here that an observed indication of the core temperature of red giants is the temperature of young white dwarfs since white dwarfs represent the denuded cores of red giants – as noted above the surface temperature of hot white dwarfs is over 25 000 K up to 70 000 K. The central star of a planetary nebula also represents the denuded core of a red giant. These stars are the hottest stars known with the highest recorded temperature for one of these stars being 200 000 K. Notice that even 200 000 K falls far short of the 100 000 000 – 4 000 000 000 K necessary for the nuclear fusion of heavy elements.

Dispersal of elements?
It is currently believed that stars in their red giant stage produce elements such as carbon, oxygen, magnesium and silicon. However, if these elements remained inside the star they would have no effect on the rest of the universe, notably in the composition of interstellar dust clouds and in the composition of planets. How the elements would escape from the interior of a star is an important question. Cassé (2003, page 158) states, "The very high temperatures required to build up complex isotopes can only occur in the deepest stellar confines, from which matter has great difficulty in extracting itself."

Several theories have been proposed to account for the dispersal of heavy elements from red giants. The main mechanism proposed is the supernova explosion. Supernova explosions are observed among Population I stars –the stars with higher metallicity. Supernovas involve the blowing a part of the core as well as the envelope of the star. This could be a means of releasing and showering space with elements hitherto locked within the core. However, the gases and debris from supernova explosions do not contain the expected elements.

Many smaller mass red giants do not explode, but simply burn out. How could their elements be released? Additional hypotheses are that red giants have thermal pulses noted as variations in their luminosity. The thermal pulses dredge up elements to the surface by convection currents (Zeilik 2002, page 374). Once at the surface, the elements are thought to be ejected into space by strong stellar winds.

Zeilik (2002, page 343) also mentions a very strong outflow of mass from the surface of a red giant in what is sometimes known as a superwind. The superwind carries off the envelope of the star which forms an expanding shell of gas heated by the remaining core. This is called a planetary nebula (because of its superficial resemblance to a planetary disk). The nebula dissipates in the interstellar medium in only a few thousand to a few hundred thousand years. The remaining core becomes a white dwarf composed mostly of carbon. IC 418, the Spirograph Nebula is an example of a planetary nebula.

5. Carbon resonance level

After hydrogen and helium, oxygen and carbon are the most abundant elements in the universe. George Gamow (1948) had proposed that these elements were produced in the first three minutes of the Big Bang. However, it became apparent that because there is no stable element with atomic mass 5 that fusion reactions had to stop at atomic mass 4 which corresponds to helium. This was frustrating for Gamow –when he discovered the presumed flaw in the nature of the light elements that prevented the heavier elements from forming, he remarked that it was "God's goof". Gingerich (2008, page 28) shows that it may not have been such a goof since the missing mass 5 may be what allows the slow burning of hydrogen in main sequence stars and hence for the stars to shine for a very long time. Nonetheless, physicists needed to find a theoretical way to cross the chasm separating helium from carbon.

The light elements lithium (Li), beryllium (Be) and boron (B) which separate helium and carbon are rare and highly unstable. They are produced in the interstellar medium as a result of collisions of highly accelerated cosmic ray nuclei of carbon and oxygen with other nuclei. Lithium, beryllium and boron are not produced in stars. Indeed, they are destroyed there as a result of their excessive fragility (Cassé 2003, page 99).

Triple alpha process

Edwin Salpeter (1952) observing the abundance of carbon in red giant stars, suggested that a triple reaction between helium nuclei otherwise known as alpha particles could produce carbon-12 in a two step process. This process is known as the triple alpha (α) process or helium fusion. It is dependent on the core of the red giant reaching 100 million K which would give sufficient kinetic energy to atomic nuclei to produce a low amount of beryllium-8.

^{4}He + ^{4}He + 95 keV (kinetic energy input) $\rightarrow$ ^{8}Be

This would be followed by a collision between beryllium and an alpha particle to produce carbon:

^{4}He + ^{8}Be $\rightarrow$ ^{12}C + 7400 keV (kinetic energy output)

(Gingerich 2008, page 28).

The problem with this reaction is that the intermediate beryllium-8 has a life time of only 10^{-16} seconds. This is a vanishingly short time before it splits again into two helium nuclei. During this lapse of time, another helium nucleus must collide with and join onto ^{8}Be for carbon to be produced. It was recognized that this was extremely unlikely –at best it would be an extremely slow process.

Resonance

Fred Hoyle noted that the red giant stars seemed to be too young to have produced the observed high abundance of carbon by such a slow process unless some physical mechanism speeded up the reaction. In 1954 Hoyle proposed that the triple alpha reaction was speeded up by an energy level a few hundred keV above the ^{8}Be and ^{4}He threshold. Nuclei have various energy levels when they are excited above the ground state. In the ^{12}C nucleus one of the energy levels is at 7654 keV above the ground state. When the energies of combining nuclei result in an energy level of the newly-formed element, resonance is achieved (Cassé 2003, page 143). Hoyle predicted from theory that this resonance level existed in carbon-12 before it had been found, and this has been taken as confirmation that the triple alpha reaction produces carbon.

This is how resonance works:
The collision of two nuclei forms a new nucleus into which is deposited the kinetic energy of the two nuclei together with the mass energy of the nuclei minus the binding energy. The new nucleus is stabilized if it can occupy one of the energy levels in its own energy scale. If the sum of the energies is not exactly right, the excess energy will be used to eject particles from the new nucleus, or if the excess is significant, this will break the nucleus in two again.

If the new nucleus comes together with the right amount of energy corresponding to one of its natural energy levels, it will be stable but descend to a lower energy level –the ground state by emitting one or more gamma photons. The correspondence between the energy of the initial particles and the appropriate level in the residual nucleus is known as resonance.

In the formation of carbon-12 the mass energies of helium-4 and beryllium-8 are fixed and cannot be altered. The kinetic energies of the reactants are dependent on temperature. Hoyle calculated the temperature from models of red giants. He predicted the existence of an energy level of carbon-12 that had not yet been found on the basis of the combined mass energies of beryllium-8 and helium-4 with the calculated temperature. The excited level of carbon-12 was found at 7654 keV above the ground state. This is just 4 % above the combined mass energies of beryllium-8 and helium-4 (Cassé 2003, pages 142-143).

Prediction and origin?
Does Hoyle's prediction really show that the origin of carbon-12 is known or was the prediction with the energy level found a coincidence? Does the production of nitrogen, oxygen and the elements in the third row of the Periodic Table show similar resonance between the mass energies of the reactants and the energy levels of the new elements? I do not have this information, but I think not or the textbooks would tell us that they do. Also, the proposed temperatures

that give rise to kinetic energy would not be so excessively high if the resonances were well adjusted.

Carbon and oxygen
Some carbon-12 is thought to convert to oxygen-16 by the following reaction:

$$^4He + {}^{12}C \rightarrow {}^{16}O + 7100 \text{ keV (kinetic energy output)}$$

(Gingerich 2008, page 29).

The combined mass energies of carbon and helium are just 1 % above an energy level of oxygen-16 (Gingerich 2008, page 29 gives it as half a percent). This means that the closest energy level in oxygen-16 is too low for the conversion of carbon-12 into oxygen-16 to proceed efficiently. Kinetic energy can always be added to the overall balance, but it cannot be subtracted. If the nearest energy level in oxygen-16 had been higher or the relevant energy level in carbon-12 lower, then all of the carbon produced would convert into oxygen, and there would be no carbon in the universe. The lack of adjustment is thought to mean that only some carbon-12 became converted into oxygen-16.

I have proposed that God created the elements of life –these are mainly the 16 or so elements at the top of the Periodic Table, including C H O N P S. If this were so, God would construct these atoms with energy levels such that they would not convert from one element into another element. We have seen that the energy levels of carbon-12 and oxygen-16 prevent carbon from transmuting readily into oxygen.

If God created carbon and the other elements of life as the components of organic molecules, He would fix the resonance levels of *all* of these elements such that they would be prevented from transforming themselves into one single type of element –for example, all ending up as iron. If each created element mutated readily into a neighbouring element, for example, if carbon transformed itself into oxygen by natural processes, then the creation of life would be rapidly undone, and there would be no life in the universe. (Although this is not to say that these elements do not have isotopes produced by natural processes).

My conclusion, controversial as it may be, is that the energy level in carbon-12 predicted by Hoyle does not prove that carbon is produced by stars, since created carbon would also have to have a series of energy levels for its preservation as carbon. The existence of this energy level does not prove the case either way since it would be necessary for both modes of origin.

In 2010 I attended a talk given by Rodney Holder, a member of Christians in Science on the subject of science and religion. Rodney Holder put forward the view that the universe appears to be fine-tuned such that life could exist in it. The Fine-Tuning argument implies the existence of a 'Fine-Tuner'. One of the main features of the Fine-Tuning argument is the fortuitous fabrication of carbon by dying stars which would convert the universe into a biofriendly place. Holder quoted Fred Hoyle, who upon discovering the predicted energy level in carbon, and although an Atheist said, "A super-intellect has monkeyed with physics, as well as with chemistry and biology and there are no blind forces worth speaking about in nature." Hoyle was saying that chance could not account for the existence of carbon in the universe.

A militant Atheist in the audience replied with force that because it has been found that resonances (or energy levels) could be different by 20 % and still produce the relevant elements, the Fine-Tuning argument breaks down; the existence of carbon could all be due to chance.

Rather than follow a Fine-Tuning argument for the production of carbon within stars, I am proposing that the energy levels found in 'fundamental elements' such as carbon *prevent* these elements from converting into different sorts of elements which would result in the disintegration of the carbon-based molecules of life.

6. Nuclear processes

I do not dispute the fundamental principles of nucleosynthesis, despite questioning the fabrication of the abundant elements of the universe inside stars. The production of new, rare elements can be both observed and reproduced experimentally using particle accelerators to bombard nuclei. Nuclear fission and nuclear fusion have also been dramatically demonstrated upon Earth.

Atom bombs and H-bombs

The atomic bomb works on the principle of nuclear fission. The fuel used for atomic bombs is ^{239}Plutonium or ^{235}Uranium. Other unstable isotopes of uranium could also be used though they are less effective. These radioactive elements naturally undergo decay by releasing neutrons. If a released neutron strikes a nucleus, it splits in two. This nuclear fission releases thermal energy, gamma rays and two or more neutrons. The escaping neutrons hit other nuclei which also split in two. If there is a critical mass of radioactive material available, a chain reaction can occur. The critical mass for ^{235}Uranium is just over a pound (0.45 kg). The chain reaction which causes the bomb to explode is produced by suddenly bringing together radioactive materials above the critical mass which have been kept separate up to the point of detonation. An atomic bomb produces thermal energy of several million K inside the exploding material itself. It also produces a fireball, an incurrent of air carrying dust into a mushroom-shaped cloud, a blast wave, lethal radiation composed of neutrons and gamma rays and radioactive debris as fallout.

The H-bomb, hydrogen bomb or thermonuclear bomb works by nuclear fusion. In this bomb, hydrogen atoms fuse to form helium in a similar way to the *P-P chain* which occurs inside stars. The fuel used for hydrogen bombs are the isotopes of hydrogen, deuterium and tritium. The fusion process is triggered by high temperatures of several million K. The hydrogen bomb is detonated by an atomic bomb to produce the required amount of heat. A nuclear weapon is composed of an atomic bomb and an H-bomb, but the most powerful component is the H-bomb. The H-bomb produces a blast shock wave which travels at supersonic speed, an intense blinding white light, intense heat producing firestorms, and it would cause worldwide radioactive fallout.

Supernova explosions

Explosions involving nuclear processes are possible on Earth, and I do not dispute that explosions also occur in the cosmos and that these produce new elements. The supernova explosions of stars produce radioactive elements and radioactive isotopes of elements. Uranium and other unstable elements are synthesized in supernovas. They decay back into lead. Gold is rare in the universe and produced in supernovas. Gold seems to be an example of an element that has hit on a stable configuration.

The elements heavier than iron have been formed during supernova explosions.

Nuclear energy

The nuclei of atoms are subject to opposing energies. The repulsion between positively charged protons in the nucleus favours smaller nuclei with lower atomic numbers (this is called the Coulomb energy). On the other hand, the surface energy of the nucleus favours large nuclei. The opposition between these energies means that the optimum mass number for a nucleus is in the middle region of around mass 60. Atomic mass 60 corresponds to the iron nucleus.

Nuclear energy can be released by the fusion of light nuclei or by the splitting of heavy nuclei in the fission process. Both these processes transform nuclei to an atomic mass closer to the middle range.

The nuclear binding energy depends on the relative numbers of neutrons and protons in a nucleus. Binding is stronger in light nuclei when the number of neutrons and protons is equal since it optimizes the neutron-proton attraction. In heavier nuclei the nuclear binding is stronger when there is an excess of neutrons, because it means that the Coulomb repulsion between protons is relatively weaker. The most stable heavy nuclei have 50 % more neutrons than protons (*Encyclopedia Britannica 2011 Standard Edition*: Atom).

Fundamental elements and linked elements

The elements of life, plus hydrogen, iron, silicon and oxygen I have labelled as the fundamental elements. These elements appear to undergo modifications producing isotopes of other elements always in groups of three. There seems to be a parent element that gives rise to the other two elements, which then decay back to the parent element.

The common, stable form of iron is ^{56}Fe, although you would expect it to be ^{52}Fe from the atomic number of 26. ^{56}Fe has four excess neutrons. Iron-56, cobalt-56 and nickel-56 are linked. It has been hypothesized that iron originates as radioactive nickel-56 produced by supernovas. The spectral analysis of a supernova observed in 1987 detected photons produced by the radioactive decay of cobalt-56. This was taken as confirmation that iron originates as nickel-56 which converts into cobalt then iron (Cassé 2003, page 152). It would also be possible that iron-56 which has four excess neutrons, converts excess neutrons into protons to produce cobalt-56 and nickel-56 under certain explosive conditions. The two products then decay back into iron over time.

Atomic number:	**Fe** 26	Co 27	Ni 28
Atomic mass:	55.845		

For examples of other linked elements see the appendix for this chapter.

Conclusion

The view I am presenting here is that elements can be divided into fundamental elements and evolved elements and isotopes. The evolved elements and isotopes came into being by natural processes such as neutron capture or nuclear fission. The fundamental elements found at the top of the Periodic Table were created as the components of life and as catalysts. The first forms of life were unicellular and were created in vast quantities together with silicate dust and iron in interstellar clouds.

The elements of life are the most abundant of the heavy elements. Heavy elements only represent 1 % of the matter in the universe compared to hydrogen and helium. The main elements of life number about 16. The 16 elements are carbon, hydrogen, oxygen, nitrogen, phosphorus, sulphur, sodium, potassium, calcium, chlorine, magnesium, manganese, iron, copper, zinc and fluorine. The fundamental elements also include iron, silicon and oxygen as silicates forming interstellar dust. There are a few other elements found in cells which are probably evolved, but for which life has found a use.

The elements which came into being as the first cells would, from then on, undergo constant recycling between organic and inorganic processes.

The fundamental elements have undergone transformations to produce the other elements found in the Periodic Table. With the exception of the inert gases, these evolved elements are mostly not very abundant. They are often unstable and radioactive.

The fundamental elements have been designed such that they will not easily transmute into other elements. If they did so, their properties would be lost and life would be wiped out. The fundamental elements can, however, be transformed in violent astronomical events such as supernova explosions. The elements produced in supernova explosions fill up the lower part of the Periodic Table. New elements have also appeared as the result of cosmic ray bombardment causing nuclear fission.

This view stands in stark contrast to the current view. The current view is that the elements of life came into existence inside red giant stars with properties ideal for life. They also came into existence in the right quantities to form life. Life then evolved from these elements by chance. The only good thing about this scenario is that it allows one to romantically say, "We are made from stardust."

It must be noted that the evolved elements generally do not have biological roles. These include the inert gases and unstable elements. Radioactive isotopes of elements do, however, become incorporated into cells along with the stable form of an element. An example of this is that radioactive carbon-14 is incorporated into living beings through photosynthesis practiced by plants along with the stable carbon-12.

In my analysis of the Periodic Table I have employed the principle that fundamental elements can be modified by natural processes. However, it is my belief that the specific properties of these fundamental elements have been fixed with the purpose of providing elements with which to form life.

Cassé (2003, page 179-180) points out that the study of galactic evolution has two aspects: the observation of abundances of elements in stars of various generations by means of spectral analysis. The other is purely theoretical, involving numerical simulation of chemical evolution. He also candidly confides,

"The problem is to harmonise an idealised model with data that are often incomplete, inaccurate and even ambiguous, in such a way as to write the universal saga of the elements, assumed to repeat itself in every galaxy. The cosmic scriptwriters propose various versions of the same tale, hoping that each will be more realistic than the others. But the conceptual framework is

94

the same. One might say that the profusion of grapeshot makes up for a certain inaccuracy in the aim."

Despite the difficulty of the subject of nuclear physics, I think there is everything still to play for.

Bibliography

Ball, Philip (2001) *Molecules: A Very Short Introduction* Oxford University Press

Cassé, Michel (2003) *Stellar Alchemy: The Celestial Origin of Atoms* Cambridge University Press

Encyclopedia Britannica 2011 Standard Edition: Aluminium; Atom; Atomic bomb; Becquerel, Henri; Bunsen, Robert Wilhelm; Chemical element; Cobalt; Copper; Cosmic rays; Curie, Marie; Curie, Pierre; Helium; Kirchhoff, Gustav Robert; Magnesium; Nickel; Secchi, Pietro Angelo; Thermonuclear bomb; Transuranium element; Zinc

Friedlander, Gerhart (1956) *Nuclear and Radiochemistry* Second edition John Wiley & Sons Inc

Frausto da Silva, J.J.R. and R.J.P. Williams (2001) *The Biological Chemistry of the Elements – The Inorganic Chemistry of Life* 2nd edition Oxford University Press

Gingerich, Owen (2008) Revisiting The Fitness of the Environment. In *Fitness of the Cosmos for Life: Biochemistry and Fine-Tuning* J. Barrow et al. eds. Cambridge University Press Pages 20-30.

McMullin, Ernan (2008) Tuning fine-tuning. . In *Fitness of the Cosmos for Life: Biochemistry and Fine-Tuning* J. Barrow et al. eds. Cambridge University Press Pages 70-94.

Sharpe, Alan G. (1992) *Inorganic Chemistry* Longman Scientific & Technical

Shaw, Andrew (2006) *Astrochemistry: From Astronomy to Astrobiology* John Wiley & Sons Ltd.

Williams, R. J. P. (1985) The Symbiosis of Metal and Protein Function *European Journal of Biochemistry* Vol. 150, pages 231-248.

Williams, R.J.P. and J.J.R. Frausto da Silva (2008) Evolution revisited by inorganic chemists. In *Fitness of the Cosmos for Life: Biochemistry and Fine-Tuning* J. Barrow et al. eds. Cambridge University Press Pages 456-489.

Zeilik, Michael (2002) *Astronomy: The Evolving Universe* 9th edition Cambridge University Press

CHAPTER 13

STAR FORMATION AND EVOLUTION

Introduction

This chapter discusses the formation of stars within bright molecular clouds, the different stages that stars pass through shining as they radiate light energy, and their eventual fate littering galaxies as cool dense objects known as stellar corpses.

Stars are composed of hydrogen at high pressure. Stars are plasmas which means that the atoms from which they were formed have been dissociated into free subatomic particles due to high temperatures. Plasma may have the consistency of water.

The question of the origin of heavy elements in the universe is linked to the nature of stars. Current theory attributes the production of the common heavy elements such as carbon, oxygen, silicon and iron to processes occurring within red giant stars. This was discussed in the last chapter. In this chapter it is proposed that interstellar dust is drawn into red giant stars, smelted and blown out again in supernova explosions.

I believe that the fundamental heavy elements originated outside stars, rather than being made inside the cores of stars. I propose that the abundant elements iron, silicon and oxygen were created as interstellar dust to serve a catalytic role in star formation.

1. Star formation in molecular clouds

Molecular clouds are clouds found within galaxies mainly composed of molecular hydrogen (H_2) associated with dark nebulas or dark clouds composed of dust. Protostars are often embedded within dark nebulas. Protostars are not visible; they are detected only as infrared sources or heat sources.

It is thought that molecular clouds and dark nebulas give birth to stars. A dark nebula of a few hundred solar masses can give birth to solar mass stars which are the commonest size of star in our galaxy. More massive stars are born on the edges of giant molecular clouds.

The Orion Nebula also called Messier 42 or M42 is a giant molecular cloud within our galaxy. The Orion Nebula contains over 500 young stars, mostly only detected in the infrared. They are being formed where molecules of hydrogen gas reach their highest density of a hundred million molecules per cubic centimetre (10^{14} molecules per cubic metre or 10^8 per cubic centimetre) and where the gas is moving at high speeds (100-150 km/second) (Zeilik 2002, page 320).

Molecular hydrogen production

Scientists interested in interstellar clouds have puzzled over how the molecular hydrogen (H_2) of molecular clouds can form from the atomic hydrogen (H) of galaxies.

Stars do not form in areas within the galaxy containing atomic hydrogen. The diffuse interstellar medium of neutral atomic hydrogen consists of clouds of low density with only about a hundred atoms per cubic centimetre. Stars do form within clouds of molecular hydrogen which have higher density. Stars are composed of hydrogen at high density.

The first question to answer in an explanation of star formation is: how does atomic hydrogen become molecular hydrogen?

It has been demonstrated that molecular hydrogen cannot form in gas phase due to the low density of atoms in the interstellar medium making collisions between atoms very few and far between. There is no realistic prospect of chemistry taking place in the interstellar medium of galaxies. Atoms which are far apart do not spontaneously join up to form bonds.

It has been suggested that interstellar dust grains may be the key to understanding. The currently accepted hypothesis is that H_2 is formed by two hydrogen atoms becoming associated on the surface of an interstellar dust grain with the dust grain absorbing the energy of bond formation. Once the stabilized H_2 molecule has formed, it desorbs from the surface of the dust grain into the interstellar medium. Thus, the silicate surface of the dust grain acts as a catalyst in bringing two hydrogen atoms into proximity such that a bond can be formed between them.

Although hydrogen molecules cannot form by the collision of hydrogen atoms in gas phase, once formed, hydrogen molecules in a gas are chemically stable since collisions between molecules are unlikely to break the covalent bond that holds the hydrogen molecule together.

Catalysts for molecular hydrogen

The problem with the current models is that interstellar dust is instrumental in forming molecular clouds; molecular clouds of hydrogen gas give birth to stars; stars form heavy elements such as silicon and oxygen by fusion reactions in their interiors; silicon and oxygen form the silicates of interstellar dust; interstellar dust is the catalyst for the formation of molecular hydrogen. Which came first? The heavy elements, the dust or the star? This is a chicken and egg problem.

Interstellar molecular clouds are continually being formed and stars are continually being born. These natural processes must have had a beginning.

As already mentioned in Chapter 11, interstellar clouds are composed of 90 % hydrogen, 9 % helium and less than 1 % heavier atoms. After hydrogen and helium, oxygen is ranked as the third most abundant element in the universe and carbon is fourth. Iron is almost as abundant as carbon. Silicon is a main constituent of interstellar dust. The Earth's crust which is composed mostly of silicate rocks contains 28 % silicon. Therefore, of the 1 % heavy elements in the universe, oxygen, silicon, iron and carbon rank high.

Iron is known to chemists as a catalyst. Iron minerals such as pyrite (FeS_2), pyrrhotite (FeS) and magnetite (Fe_2O_3) have active charged surfaces capable of promoting chemical reactions. They have hydrogen evolving reactions on mineral surfaces in water (Shaw 2006, page 251). Powdered iron is used as a catalyst to react together hydrogen and nitrogen in the production of ammonia in the Haber process.

Interstellar dust grains are composed mostly of silicates. Clay which is a silicate mineral is known to have catalytic properties.

Interstellar dust grains tend to be about one micron (1 μm) in diameter, although some interstellar dust grains are much larger than this. An example of small sized dust grains are those in the circumstellar disk around the star β Pictoris (Piirola et al. 1992, page 400). Many dust grains measure only a fraction of a micron and there are many more small grains than large ones (*Encyclopædia Britannica 2011 Standard Edition*: Nebula). The small size of these interstellar dust grains would make them good catalysts since having small size, they would present a large surface area.

In Chapter 12 I suggested that the Periodic Table of Elements could be divided into fundamental elements at the top of the chart and evolved elements lower down in the chart. I proposed that about 12 of the fundamental elements were created with specific properties to serve as components of life. I mentioned briefly that three of the fundamental elements would have been created as catalysts.

I propose that the elements iron, silicon and oxygen were created as dust in the fraction of a micron size range within the atomic hydrogen medium of early galaxies. These iron and silicate catalysts allowed the formation of molecular hydrogen from atomic hydrogen to take place, giving rise to molecular clouds. The first stars were born from these molecular clouds by natural processes. Therefore, I propose that the natural processes that continue to produce stars today were originally set up by the creation of elements to serve as catalysts.

The **Iron and Silicate Catalysts Hypothesis** can be stated thus:

The elements iron, silicon and oxygen were created as iron and silicate catalysts in the form of fine interstellar dust within the atomic hydrogen gas of galaxies. The catalytic properties of these elements allowed atomic hydrogen to be converted into molecular hydrogen forming molecular clouds. Stars form within molecular clouds by natural processes.

The notion of the creation of fundamental elements acting as catalysts in the formation of molecular clouds cracks the chicken and egg problem.

2. Molecule formation and rotation in stars

According to the current models, protostars are formed by gravitational collapse within molecular clouds. "A cloud with enough mass and a low temperature will naturally contract from its own gravity." (Zeilik 2002, page 317). In these models collapse occurs and matter goes into free-fall onto itself forming a dense core that becomes a star. However, the models do not explain the observed rotation of stars; the adding in of angular momentum shows up problems with these theoretical models. Zeilik (2002, page 318) writes,

"The addition of spin to theoretical models of protostar collapse makes the calculations much tougher and the results less conclusive. To give you the flavor of results: in some cases, a ring or bar of material results. These rings and bars turn out to be unstable in some instances; they break up into two or three blobs, which sometimes coalesce."

Another idea is that a supernova explosion gives rise to a shock wave that as it moves through a molecular cloud compresses material together. The compressed material becomes dense enough for a star to form. Pressure waves could also be produced when galaxies collide. In support of this idea there is evidence that stars form in great numbers within a galaxy after a collision.

I propose that molecule formation itself within the molecular cloud gives rise to star formation. When molecules form, chemical bonds are made with the sharing of electrons; this holds two or more atoms together. This could be conceived as a 'clumping of matter' or contraction of space at the nanoscale. When this tiny effect is multiplied millions of times over by the formation of large numbers of molecules it could cause a large-scale effect.

Stars rotate because the molecular clouds from which they were formed are rotating. Where did this angular momentum come from? I suggest that motion within a molecular cloud is fuelled by the energy released from bond formation. By this I mean that the chemical heat energy released when hydrogen atoms join to form molecules would be converted into kinetic energy fuelling the rotation of the cloud. The conversion of chemical energy into kinetic energy within the cloud would have the effect of carrying away the excess energy and stabilizing the molecules formed so that they would not break apart again. It seems that hydrogen molecules are formed on the surfaces of fine calibre interstellar dust grains; it is this dust that would set the cloud in motion.

In general galaxies are also rotating. The origin of their angular momentum will be discussed in Chapters 15 and 16. The motion in a particular direction of clouds of atomic hydrogen belonging to the spiral arms of a galaxy may confer direction upon the rotation of the newly forming molecular clouds.

According to current theory, star formation occurs by gravitational collapse in molecular clouds. In the view presented here, molecule formation rather than gravitational collapse initiates star formation.

Interstellar dust composed of silicates and fine calibre iron, although they only represent 2 % of a dark cloud, play a vital role in the birth of stars. Natural processes occurring at the nanoscale en masse contribute to a macroscale phenomenon. It is the formation of molecular hydrogen itself which causes an increase in density that becomes the seed of a new star.

Once a core of dense matter has formed within a molecular cloud and reached a critical mass, a protostar may continue to grow by accreting matter by gravitational attraction. As the star accumulates matter, gravitation may take over as the main force behind further growth of the star.

3. Stellar evolution

This section is a summary of the accepted classification of stars.

The word 'evolution' in this context denotes the idea that stars pass through a series of phases through time on a cosmic scale; they undergo change. The word 'evolution' is not used for the origin of stars; stars are said to form or be born within molecular clouds, not 'evolve' in terms of their beginning or origin. It is true to say that this use of the word 'evolution' is similar to the 19th century use of the word 'evolution' in the context of biology. Nineteenth century biologists such as Richard Owen and Louis Agassiz used the word 'evolution' to signify the development of an organism through different stages as an embryo before birth and then a juvenile and then an adult. One could say that stars undergo 'development' rather than 'stellar evolution', although there is no 'blue-print' to stellar development as there is behind biological development. Stars undergo a sequence of stages marked by particular physical processes which depend upon their size, temperature and age.

A star like the Sun will live for ten thousand million years whereas more massive stars live only for a few million years. The reason for this is that massive stars burn their fuel (hydrogen) at a greater speed. The fact that we observe massive stars within galaxies now shows that stars must continue to be born. If stars had only been formed at one time, the larger ones would not be observed now.

Nobody has ever observed a star evolve, but observation of the types of stars in the galaxy indicates that stars pass through a number of stages from newly formed protostars to main sequence stars, ending their lives as stellar corpses. This evolution occurs by natural processes which stellar models attempt to describe.

Protostar stage
Protostars are born in molecular clouds and they are detected in the infrared as heat sources. They cannot be visibly seen at this stage. Once a core has formed, it accretes material from a surrounding envelope of material by gravity, and heats up.

Molecules of hydrogen gas are converted into atoms and release radiation as the temperature approaches 4000 K. The protostar continues to draw material in, and becomes large in size and luminous. The young star is shrouded in dust which extinguishes its light from outside view.

After about a million years, the hidden star finally emerges from its womb of dust and becomes visible either by accreting the dust cloud onto itself or by dissipating the clouds of dust (Zeilik 2002, page 317). This pre-main sequence star is luminous, although larger and cooler than it will be when it reaches main sequence.

Main sequence stage

After some 50 million years a solar mass star joins the main sequence stage. The term main sequence comes from the Hertzsprung-Russell diagram or H-R diagram which shows the surface temperature of stars along the horizontal and luminosity on the vertical axis. The main sequence is stars forming a diagonal line on the diagram. This series of stars has decreasing mass, but similar chemical composition.

13.1 This is an approximation to the
HERTZSPRUNG-RUSSELL DIAGRAM

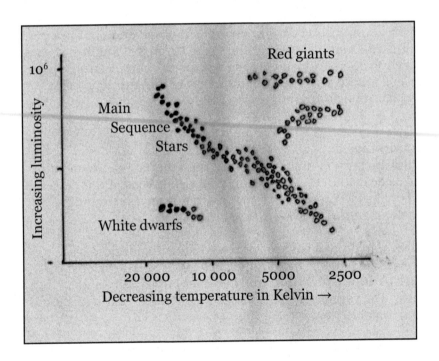

Towards the top left corner of the diagram are high mass stars which are hotter and more luminous, and towards the bottom right corner are low mass stars which are cooler and less luminous. The surface temperature of main sequence stars varies between 30 000 and 3500 K. The main-sequence lifetime for a star like the Sun represents about 80 % of its total lifetime.

During the early star phase the core has been contracting and heating up due to gravity. When the core reaches 8 million K thermonuclear reactions are turned on inside the star and it enters main sequence. *P-P chain* reactions or hydrogen fusion converts hydrogen into helium (H $\rightarrow$ He). The star shines as matter becomes energy. When nuclear fusion starts, the star stops contracting because the inward pressure of gravity is counter-balanced by the outward pressure of a hot gas and radiation. The rate of energy radiating from the surface of the star equals the rate of energy production in the core. In main sequence stars the massive inward forces of gravity are counteracted by the outward pressure of fusion reactions and the forces are held in balance so that the star neither expands nor contracts.

The early stage when the star continually grows in mass due to matter being drawn in ends when the flow of matter is reversed. When the nuclear furnace is ignited, strong stellar winds blow out from the star's surface. Stellar winds are a feature of main sequence stars and may be the means by which shining stars have cleared away the dust clouds from which they were born.

In the case of our Sun, the stellar wind is known as the solar wind. The solar wind consists mostly of protons and electrons accelerated by the high temperature of the solar corona (outer layers of the Sun). The solar wind travels out to a distance of 20 AU[3]. Planets deflect the solar wind by magnetic fields. The solar wind blows most particles smaller than 10^{-13} gram out of the solar system (Zeilik 2002, page 337). The solar wind is greatly increased during solar flares. Solar flares emanate from Sun spots which are Earth-sized pores in the Sun's surface.

Main sequence stars emit ultraviolet light as well as visible light. The UV radiation ionizes hydrogen by knocking out electrons to leave charged hydrogen nuclei. This leads to circumstellar zones of ionized hydrogen atoms called H II regions where the ionized hydrogen may have a temperature of 10 000 K.

Stars are plasmas. Plasma is formed when heat peels off the electron shells from atoms leaving completely ionized positively charged nuclei and free electrons. In an atomic gas the electrons are held captive by atomic nuclei. In a plasma gas the atoms are totally dissociated by

---

[3] AU stands for Astronomical Unit. It is the distance between the centre of the Sun and the Earth which is about 150 million kilometres.

heat and can have high density. The plasma of the Sun has the density of water and the free movement of electrons makes it a good electrical conductor.

The main sequence stage ends when the core has almost entirely been converted to helium. Massive stars burn hydrogen at a much faster rate than low mass stars. Although massive stars start with much greater supplies of hydrogen from the outset, they exhaust them quicker. A 15-solar mass star will have luminosity 30 000 times greater than the Sun, but live for only 5 million years.

Red giant stage
When the core of a star has been converted to helium the *P-P reaction* of hydrogen fusion ceases. At the red giant stage the star expands a hundred times or more larger than before. This expansion in size causes the surface temperature to drop, although the luminosity increases due to the large surface area from which to radiate. The star appears redder, hence the name red giant.

Once fusion reactions cease the star cools off. The surface temperature at this stage is between 6000 and 2000 K.

Red giants are associated with silicon carbide (SiC) grains of various sorts.

Current theory postulates that red giants enter a phase of helium burning to produce carbon by the triple alpha process in the core. I question this and offer a different explanation for the presence of carbon in red giants in the next section.

Red giants flash and pulsate. They undergo various fates depending on their size. Some undergo a series of bursts that blow off the outer layers to form a planetary nebula, while the more massive red giants end in supernova explosion.

Planetary nebula
Stars of one to ten solar masses may go through the phase of planetary nebula (Shaw 2006, page 97). The red giant develops a very strong outflow of mass from its surface. This has been called a superwind, but it could be better described as a series of mini-explosions. Over

106

about a thousand years the outbursts rip off the envelope of the star, leaving behind a hot core which appears as a central star.

The expelled material forms an expanding shell of bright gas known as a planetary nebula. The series of bursts may form concentric shells of material typically expanding at 20 km/second. The gas may have a temperature of 10 000 K and fluorescent regions contain ionized elements. This nebula (which has no connection with planets) expands outwards until it dissipates into the interstellar medium after about 50 000 years.

The denuded core of a planetary nebula becomes a white dwarf. An example of a planetary nebula is the Spirograph Nebula IC418.

White dwarfs

Red giants end their life in explosion. A nova is an explosion that blows off the outer layers of a star, while a supernova is an explosion that blows apart the interior regions also (more on novas and supernovas later). After nova and maybe supernova explosions have occurred, a core is left behind where fusion reactions have ceased. At this stage gravity takes over.

One scenario is that the core becomes a white dwarf. This is when the core implodes and may become the size of the Earth although it has the mass of the Sun. White dwarf stars are still hot and emit radiation, but they are very small in size. The small size means that they have low luminosity which makes them difficult to see. A hot white dwarf has a surface temperature of over 25 000 K, while a cool white dwarf has a surface temperature of 4000 K or lower. Over time this remnant core cools down.

When a star no longer burns fuel the hot gas pressure of the plasma no longer resists gravity and gravity crushes the star. However, the free electrons form a degenerate gas while the atomic nuclei form a crystal structure embedded in the degenerate gas. At this stage electron degeneracy pressure holds up the white dwarf star keeping it stable at a higher density than a main sequence star.

Many white dwarfs contain a mixture of mainly carbon and oxygen (Zeilik 2002, page 360). When a white dwarf of this sort has cooled down it is thought to form a solid carbon crystal (see footnote[4]).

Neutron stars

A star larger than the Sun (estimates of size vary), when it has undergone supernova, will end its life by the core imploding to an even smaller size than a white dwarf –the neutron star has a diameter of only 10 – 20 km. It is composed of a degenerate gas which means that the more mass it has, the smaller its diameter will be.

The crushing process of gravity is greater in neutron stars than in white dwarfs. As the neutron star implodes and reaches higher and higher density, at a certain point, inverse beta decay occurs whereby an electron, also called a *beta particle* and a proton combine to form a neutron and a neutrino. At great pressure neutrons are no longer bound to nuclei and start to form a degenerate gas. This gas will finally contain 80 % neutrons and a spoonful of it, if weighed on Earth would weigh a thousand million tons (Zeilik 2002, page 361).

The neutrons provide a neutron degeneracy pressure which balances the inward pull of gravity, and a stable neutron star is formed.

Pulsars are small, dense, rapidly rotating objects. It is thought that pulsars may be neutron stars. Pulsars emit a beam of radiation composed of accelerated electrons. Only an extremely small object could rotate in milliseconds as pulsars do. Neutron stars fit the bill since they measure only about 10-20 km across. The pulsar PSR 1937+214 spins at 642 times per second and its surface rotates at about a tenth of the speed of light.

---

[4] The ancient Greek cosmologists proposed that the heavenly bodies, unlike the Earth, were perfect and made of shining crystal. Maybe they were not so far wrong after all since white dwarfs could be likened to shining diamonds in the sky.

Black holes

A black hole may be left after a supernova explosion and the collapse of the core of a star above three solar masses. When the density of the core reaches a certain critical value, photons emitted by the star are trapped within the intense gravitational field. Instead of moving away in straight lines, the light waves start to orbit the star because the escape speed from the star's surface is greater than the speed of light. Since no light escapes from it, the star is known as a black hole.

The former view and my own view is that stellar corpses named black holes are extremely dense, solid objects probably composed of a tight ball of neutrons. The Singularity View is that black holes really are holes in spacetime (this will be discussed in Chapter 15, section 6).

Black holes are very small in size —only a few kilometres across and they do not emit light. This means that black holes cannot be seen and they are difficult to detect. However, black holes are known to exist by the effects their enormous gravitational fields have on surrounding matter. In the constellation Cygnus there is a binary star system consisting of a blue supergiant star and an invisible companion that revolve around each other with a period of 5.6 days. This binary emits X-rays that are thought to come from matter from the supergiant star being sucked into the companion black hole. The infalling matter is accelerated and intensely heated causing it to emit X-rays before disappearing into the black hole. This black hole is labelled Cygnus X-1.

According to the program 'Stephen Hawking's Universe' 3/3 broadcast on Channel 4 on the 2nd October 2010 black holes crush down to 12 miles in diameter, the size of a city. However, there are supermassive black holes at the centres of galaxies. There may be a supermassive black hole at the centre of our galaxy equivalent to the mass of four million Suns with a diameter of 11 million miles.

Black dwarfs

When a white dwarf has finally cooled down and emits no light it becomes a black dwarf. This process takes several thousands of millions of years. The black dwarf cannot be seen, although its mass could be felt by certain gravitational effects it produces. It is the final stellar corpse.

4. Dust drawn into stars

Some cool red giants and supergiants are enveloped in circumstellar material containing both gas and dust. Red giant stars have been called carbon stars since they have atmospheric carbon. Some supergiant stars have spectra showing the presence of silicates around them. Carbon is more abundant than oxygen in a few red giants which indicates the presence of graphite particles (C) or silicon carbide (SiC) grains.

Current theory presents red giants and supergiants as furnaces producing a whole series of elements including carbon, oxygen and silicon (this was discussed in Chapter 12). The elements are said to be blown out of red giants by strong stellar winds or escape the interior of the star by supernova explosion. These elements would then form interstellar dust clouds which produce protostars.

Red giants are stars whose surface temperature has dropped enormously due to the huge increase in their size producing a very large surface area. The enlargement of red giants is explained by an increase in their opacity. The argument is as follows: stars transport energy from the core to the surface by three basic methods; conduction, convection and radiation. If a star's material is more opaque, radiation will flow through it much slower. Opaque material will act like an insulation jacket around a star. As the opacity of a star increases, the radiation becomes dammed up; this causes the star to become hotter inside with higher pressure, and the star expands. At this stage the expansion of the star is due to convection taking over from radiation to transport energy to the surface of the star.

The increased opacity of a star is caused by an increase in the star's size and a decrease in its surface temperature. "The temperature drop increases the opacity, and convection carries the energy outward in the star's envelope." (Zeilik 2002, page 342). If this were the only criteria, it would involve circular reasoning. The opacity of a star also depends on its chemical composition. A greater abundance of metals (or heavy elements) increases opacity (Zeilik 2002, page 337).

I am going to describe a model that gives a different explanation for the association between red giants and the elements carbon and silicon:

- When a star runs out of hydrogen fuel in its core, fusion reactions producing helium cease and it comes off the main sequence. At this stage the stellar wind ceases, UV radiation is not produced as before and the star is cooling down.

- Dust from surrounding interstellar clouds is drawn towards the dying star since it is no longer kept away by stellar wind.

- Life containing carbon proliferates in association with the dust since it is not killed by ultraviolet radiation or extreme heat. The dust grains associated with life form circumstellar rings around the cooler star.

- The favourable conditions for unicellular life bring about large accumulations of dust rings around the red giant which gravity starts to draw into the star.

- The dust composed mainly of silicates and carbon molecules, drawn into the surface layers of the star, raise the opacity of the star.

- The insulating effect of dust which is composed of heavy elements obstructs the loss of heat from the star by radiation, and the star expands.

- The enormous expansion of the red giant engulfs the circumstellar clouds of dust surrounding it which increases the carbon and silicate components of its surface layers.

In this view, elements such as carbon, oxygen and silicon flow into or are engulfed by the red giant causing it to become opaque and enter a phase of massive expansion. In this scenario, lack of a stellar wind allows dust to be drawn into the star. On the contrary, according to the current theory, the elements move in the other direction, out of the red giant star due to a very strong stellar wind.

According to the currently accepted theory of nucleosynthesis within red giant stars (using *Encyclopædia Britannica 2011 Standard Edition*: Chemical element as a reference):

- Carbon and oxygen are produced by helium burning at 100-200 million K

- Magnesium, sodium, silicon and sulphur are produced by carbon and oxygen burning at 500-1000 million K

- Chromium, manganese, iron, cobalt and nickel are produced by silicon burning at temperatures up to 4000 million K

(Helium is produced by hydrogen fusion in main sequence stars at 8 million K).

The comment that I would like to make is that if carbon and silicon can be dredged up from the core of a red giant by a strong stellar wind such that these elements are detected in the star's outer atmosphere, why is the supposed heat of the core not also released to the surface of the star? If there are processes which can release newly-made elements from the interior of a star, how can the core have a temperature of between 100 million and 4000 million K, while the surface has a temperature of only between 2000 and 6000 K? There is no observed evidence of temperatures existing above several million K in the universe. The temperatures given for helium, carbon and silicon fusion have been calculated from the amount of energy required to overcome the repulsive force of protons in nuclei if new elements were to be made in this way. Nobody has ever seen inside a star. If the surfaces of stars show no evidence of stars being this hot, these temperatures are just 'theory-generated facts' required to make the theory work.

Stellar convulsions

Red giants undergo various types of convulsions. They are observed to flash and pulsate. A red giant that forms a planetary nebula may expel its outer layers by a series of bursts. Nova and supernova explosions have already been mentioned, the nova involving the outer layers of the star while the supernova blows apart the whole star. These seem to be different degrees of the same phenomenon.

I suggest that red giants draw in silicate dust, organic carbon molecules, water as ice associated with dust grains and finely divided iron (which is abundant in interstellar clouds).

If large amounts of dust associated with ice were drawn into a red giant, the sudden vaporization of water could have explosive results. The vaporization of water is one of the causes of volcanic eruptions on Earth.

If the red giant came from a main sequence star of considerable age, the main sequence star could have had a planetary system in orbit around it. At the red giant stage, the massive enlargement of the star to the size of a 'solar system' would engulf any planets orbiting within an inner planetary system. The enveloped planets would be vaporized inside the star, and could cause it some digestive problems and result in explosive bursts.

The molecules drawn into a star would be dissociated into ionized atoms by the heat of the star. Iron drawn into the star would be smelted in the outer layers of the star, and possibly transformed into radioactive cobalt and nickel in the hotter interior of the star. It may be that at a certain threshold in the quantity of material, nuclear fission of unstable radioactive elements produced within the red giant star would produce an explosion similar to that of an atomic bomb.

The nuclear fission of radioactive materials releases neutrons, thermal energy and gamma rays in a chain reaction. On Earth the detonation of atomic bombs has produced exploding material with a temperature of several million K. The exploding material of a red giant in a supernova explosion may transform iron into other elements (found lower in the Periodic Table of Elements); these include the radioactive element uranium.

Novas and supernovas

Novas are detected when a star increases in brightness in a rapid space of time of just a few days and stays at peak brightness for a few hours. When this happens the star's outer layers expand from 100 to 300 solar radii and then collapse back onto the star. This blows off a shell of material from the star that rapidly expands away from it. It is calculated that about 100 novas take place in our galaxy, the Milky Way every year. The Firework Nebula (GK Per) is an example of the remnant of a nova with expansion of up to 1200 km/second (Zeilik 2002, page 363).

Nebula material produced by novas show enhanced abundances of carbon, nitrogen, oxygen and neon.

Supernovas occur in more massive stars. They are cataclysmic events which may be 100 000 times more luminous than a nova; they may even shine with the brightness of an entire galaxy. Over the age of our galaxy hundreds of millions of supernovas must have taken place, although only six have been recorded in history. Hundreds of supernovas have been observed in other galaxies. The Crab Nebula is a supernova remnant in our galaxy.

Type I and type II supernovas have different durations, type I dying away more gradually. Type I supernovas occur in Population I and II stars. Type II supernovas are associated only with Population I stars which are more recent, younger stars that are richer in metals. (Populations of stars will be described in Chapter 16). The spectra of Type II supernovas show strong hydrogen lines showing the presence of hydrogen. The spectra of Type I supernovas show iron, nickel and cobalt. (Zeilik 2002, pages 366-367).

Type I supernovas appear often to occur in one-solar-mass stars. It is not understood how a solar-mass star could explode with the violence of a supernova. A model involving binary stars is used as an explanation in Zeilik (2002, pages 366-367). Denton (1998, page 198) suggests that it is the iron in the centre of a star which triggers a supernova explosion. I suggest that iron is drawn into a red giant from the interstellar medium; while inside the star iron is transformed into radioactive elements; the accumulation of radioactive elements eventually destabilizes the star; at a certain point the star explodes like an atomic bomb showering space with radioactive elements produced inside the star and during the explosion itself.

In continuation with this line of thinking, I have formulated a further hypothesis in which the iron drawn into red giants is smelted into molten globules of iron. When supernova explosion occurs these molten lumps of iron are shot out. The silicates of interstellar dust drawn into the star are also melted into lumps of rock and shot out. This will be the subject of Chapters 17 and 18 in Part III).

5. Failed stars

In this chapter we have looked at all the different types of stars. To complete the picture, we must now examine stars that started out so small that they never made it as stars. They are the brown dwarfs. They may have started out as red dwarfs – red dwarfs are hotter than brown dwarfs. There is a strong possibility that they fade and shrink into giant gaseous planets.

Brown dwarfs and red dwarfs

Brown dwarfs are very small stars with insufficient mass to ignite fusion reactions in their core. They remain as protostars, never entering the main sequence. Brown dwarfs emit only red light and heat as infrared radiation. Young brown dwarfs, in their first thousand million years, can have temperatures and luminosities similar to some stars at this stage. The surface temperature of the Sun in its protostar stage before nuclear reactions started up would have been about 3500 K. Brown dwarfs cool down to under 3000 K, and eventually approach planet-like temperatures of under 1000 K.

Under one classification, brown dwarfs range in size from the lowest mass stars down to the mass of a giant gaseous planet such as Jupiter; the largest brown dwarf would be 75 times the mass of Jupiter. There is debate as to the lower size limit of brown dwarfs. Under another classification, stars that range in mass from smaller than the Sun to a tenth of the mass of the Sun are classified as red dwarfs, while brown dwarfs range from the smallest red dwarf to the size of Jupiter. The label red dwarf is understandable since these small stars emit red light. Red dwarfs are very common.

Brown dwarf stars are difficult to detect, but in 1995 a brown dwarf called Gliese 229B was identified as the companion of a red dwarf star, Gliese 229A. Gliese 229B has a mass about 40 times greater than Jupiter and a surface temperature below 1000 K.

Jovian planets

Brown dwarfs or red dwarfs are failed stars. They were too small for a nuclear furnace to ignite and bring them to true stardom. They produce heat and low light, but they also cool rapidly and shrink.

They take on the aspect of gaseous planets, also known as Jovian planets in both their temperature and composition.

Stars and planets form in different ways. Stars are formed in interstellar molecular clouds. The planets of our solar system are all supposed, according to the Nebular Theory, to have formed from a solar nebula. Encyclopedia Britannica states, "Similar in appearance to high-mass planets, brown dwarfs may be distinguishable from planets only in their formation mechanism" (*Encyclopædia Britannica 2011 Standard Edition*: Brown dwarf).

I propose that the giant gaseous planets such as Jupiter and Saturn, Uranus and Neptune are, in fact, brown dwarfs and that they originally formed as stars. In Chapter 17 of Part III I contest the Nebular Theory of planet formation, and offer an alternative explanation showing how a solar system could be formed of diverse planets of very different composition, some gaseous and some terrestrial. According to the hypothesis presented in Chapter 17 some of the planets in our solar system could be failed stars.

Jupiter, the largest body in the solar system after the Sun, has various brown dwarf characteristics. It is composed mostly of hydrogen and helium, the upper atmosphere being composed of 82 % hydrogen and 18 % helium. Jupiter radiates about twice the energy into space as it receives from the Sun. The composition of the core is unknown, but it is estimated that the core may have a temperature of about 11 000 K with an inner core temperature of 40 000 K, while the surface temperature is only 130 K (Zeilik 2002, page 203). Saturn is also large in size, though smaller than Jupiter, composed of hydrogen and helium, and emits more infrared radiation than it receives from the Sun. Saturn's core may have a temperature of 10 000 K with an inner temperature of 15 000 K.

Neptune is a cold, icy planet at 60 K, but if it were heated by the Sun alone it would be at 44 K. Neptune has internal heat and gives off three times as much heat as it receives from the Sun. Uranus also has internal heat.

Some astronomers believe that there is a continuum between small brown dwarfs and large gaseous planets since they have the same composition of mainly hydrogen and helium and they emit energy. I espouse this minority view.

Conclusion

In this chapter I have presented the hypothesis that fine silicate dust made from silicon and oxygen, and fine calibre iron were created as catalysts that would transform the atomic hydrogen of galaxies into the molecular hydrogen of molecular clouds. This is the Iron and Silicate Catalysts Hypothesis involving nanocreation. Stars are continually born in molecular clouds by natural processes.

I have linked the formation of stars with molecule formation by proposing that molecule formation brings about a contraction of space and therefore a rise in the density of matter within molecular clouds. At the same time the chemical energy released by molecule formation is converted into kinetic energy – the energy of motion driving the rotation of molecular clouds. This angular momentum is transferred to the rotation of stars.

Most of the chapter describes stellar evolution which involves the different stages that stars go through between their birth as protostars and death as stellar corpses.

In section 4, I focused on the connection between red giant stars and heavy elements such as carbon. I propose that the carbon and silicates observed in the outer layers of these dying stars come from the drawing-in of interstellar dust rings by the star in the absence of stellar winds.

I propose that it is the difficult digestion of heavy elements drawn-in that transforms a dying star into an enlarged red giant. The convulsions of red giants may be provoked by the engulfing of planetary systems when the star expands. Destabilization of the star may occur due to the vaporization of water ice and the production of unstable radioactive elements within the star. The nova and supernova explosions to which red giants are prone may be caused by the fission of radioactive elements within the star in a mode similar to the detonation of an atomic bomb.

The chapter ends with small, cool objects which failed to reach stardom. They are the brown dwarfs which appear to cool down and shrink until they leave only a small remnant. The gaseous planets of our solar system may be failed star remnants.

Bibliography

Cassé, Michel (2003) *Stellar Alchemy: The Celestial Origin of Atoms* Cambridge University Press

Encyclopædia Britannica 2011 Standard Edition: Black hole; Brown dwarf; Chemical element; Nebula; Solar wind; Sun.

Piirola, V., F. Scaltriti & G.V. Coyne (1992) Circumstellar disks deduced from sub-arcsecond polarization observations of two young stars. *Nature* Vol. 359, pages 399-401.

Shaw, Andrew (2006) *Astrochemistry: From Astronomy to Astrobiology* John Wiley & Sons Ltd.

Taylor, Stuart Ross (1998) *Destiny or Chance: Our Solar System and its Place in the Cosmos* Cambridge University Press

Zeilik, Michael (2002) *Astronomy: The Evolving Universe* 9th edition Cambridge University Press

CHAPTER 14

THE STRUCTURE OF MATTER

Introduction

In this chapter we are going to look at the fundamental composition of matter. This is the subject matter of particle physics.

This brief chapter is mainly about what is known about matter at the subatomic level. Hypothetical particles and theories that try to explain the cosmos in terms of unknown laws and forces will be mentioned, but not described at length; these theories play no part in my own thesis.

I believe that God created matter; God created matter in such a way that its properties allow matter to interact and produce a stable but evolving universe with duration through time.

This chapter serves as an introduction to the next chapter which deals with the explosive start to the macrostructures of the universe starting from its subatomic components.

1. The atom

The structure of atoms and the way in which they interact is fundamental to the working of the universe. Each element is a type of atom. The simplest element in terms of atomic structure is hydrogen; it is also the most abundant element in the universe.

Hydrogen
The hydrogen atom consists of a nucleus composed of one proton with a positive electrical charge, orbited by one electron with a negative charge. The positive and negative charges equal each other so the atom does not carry a charge. Neutral hydrogen is composed of hydrogen atoms. A hydrogen ion is a proton with no electron. It is a charged particle. Ionized hydrogen consists of protons alone. When two hydrogen atoms combine to form a molecule of hydrogen, the two

electrons are delocalized and shared by both protons. The electrons follow paths around both nuclei and this constitutes a covalent chemical bond. Molecular clouds are composed mainly of molecules of hydrogen.

Neutrons

Apart from hydrogen all other atoms have more than one proton in the nucleus. Being positively charged protons repel each other so this is only possible because of the presence of another particle called a neutron. The neutron is an uncharged particle that adds to the attractive force that holds the nucleus together, but does not affect the repulsive force. A neutron is formed from the combination of a proton with an electron. Deuterium is a form of heavy hydrogen. It has a nucleus with one proton and one neutron orbited by one electron.

Thus, atoms have a nucleus consisting of massive particles that are protons and a more or less equal number of neutrons (different isotopes of an element have different numbers of neutrons). Protons and neutrons are about 2000 times more massive than electrons. Protons and neutrons are collectively known as nucleons. The nucleus is orbited by electrons that are tiny particles equal in number to the number of protons in the nucleus. In natural elements (of any significant abundance) this number ranges from 1 (hydrogen) to 92 (uranium). The size of the atom depends on the number of particles it contains.

Empty space

The atom could be likened to a miniature solar system except that the electrons orbit the nucleus in all directions and at all angles. The orbitals of electrons are not always circular, but when the orbital of one electron is elongated in a certain direction, the orbitals of other electrons are elongated in other directions to achieve an overall spherical shape. Not all nuclei are spherical; some are shaped like a lemon. Such deformed nuclei result from shells in the nucleus being only partially filled by nucleons.

The distance from the nucleus to the closest position of an electron – the first electron shell is about a thousand times the diameter of the nucleus (Zeilik 2002, page 360). However, electrons may be

120

positioned much further out than this. The result is that the atom is mostly empty space. Mader (2007, page 20)[5] gives the following analogy:

"If an atom could be drawn the size of a football field, the nucleus would be like a gumball in the center of the field, and the electrons would be tiny specks whirling about in the upper stands. Most of the atom is empty space. We should also realize that we can only indicate where the electrons are expected to be most of the time. In our analogy, the electrons might very well stray outside the stadium at times."

Or;

"If the proton of a hydrogen atom were the size of a marble, the electron would be about a kilometer away from the nucleus." (Zeilik 2002, page 93).

Energy states –electrons

Atoms have different energy states. An energy state is the size of the electron orbit –the larger the orbit, the higher the energy state. If an atom absorbs energy from light, its electrons will be moved outwards into a larger orbit around the nucleus. If the atom emits light, its electrons will take up orbits closer to the nucleus. When atoms emit light they move from an excited state back to the ground state. The ground state is the normal state of an atom. It represents the state with the least energy with the electrons tightly bound to the nucleus. Consequently, when atoms absorb energy they move into excited states, and when they make the transition to a lower energy state they emit light.

---

[5] Mader, Sylvia S. (2007) *Biology* 9th edition McGraw-Hill

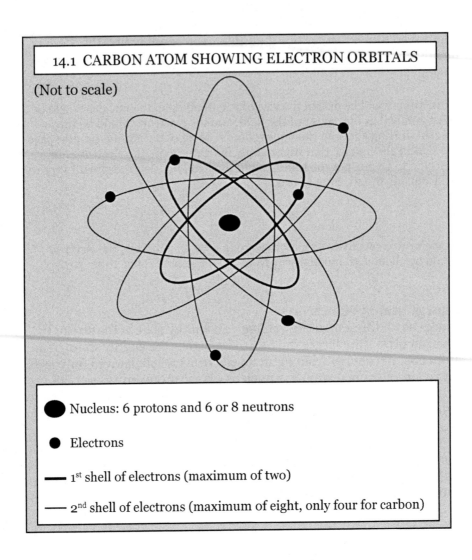

14.1 CARBON ATOM SHOWING ELECTRON ORBITALS

(Not to scale)

⬤ Nucleus: 6 protons and 6 or 8 neutrons

● Electrons

━━ 1st shell of electrons (maximum of two)

━━ 2nd shell of electrons (maximum of eight, only four for carbon)

Quantum Theory

There were two famous contributions to the understanding of light. James Clerk Maxwell (1831-1879) a Scottish physicist proposed the electromagnetic theory in which radiation travels at the speed of light in waves. He published four fundamental equations to explain electromagnetic phenomena in 1873. Waves are undulations in intertwined electric and magnetic fields. At the turn of the century Max Planck (1858-1947) a German physicist proposed that radiation not only travels as waves, but also as streams of packages of energy that behave as particles now known as photons. This is known as the theory of Quantum Electrodynamics (QED).

Quantum Theory explains that light is absorbed and emitted by atoms only in discreet packages called quanta. A quantum of light is called a photon. The mechanics of the atom are such that electrons orbiting the nucleus remain in stable states of specific energy levels unless quanta of energy are absorbed moving electrons into larger orbits and higher energy levels. When an excited atom falls back to its ground state it emits photons which are parcels of energy of fixed value and specific frequency. The fixed value or quantum number represents the angular momentum or angular velocity multiplied by mass of the orbiting electron.

Thus, the energy quanta of light can be transferred into the kinetic energy of electrons. For example, the electron in a hydrogen atom can jump up to a larger orbit, if it is given an energy boost. However, the electron must absorb energy from a photon with exactly the right energy to do so. Only a particular frequency of light will raise the electron's energy level.

Knocking out electrons
There are various consequences to the interaction of photons with electrons in the atom. Photons with very high energy can knock an electron beyond the outer orbit such that it leaves the atom. When this occurs the atom is ionized. Different elements consist of atoms with varying numbers of electrons orbiting in successive electron shells around the nucleus. The electron shells are numbered 1 to 7 on the rows of the Periodic Table. It is the electrons in the outer-most electron shell that may be lost.

When atoms bind together to form molecules, it reduces the total energy and for this reason they remain bound together once formed. The shared electrons of molecules can change their disposition by absorbing light energy. Strong light causes many molecules to dissociate into individual atoms.

Collision between atoms

When atoms absorb photons they become excited. Photons are only absorbed if they carry the energy corresponding to the difference in energies between two energy levels of the atom. Atoms also become excited when they obtain energy through collision. When a gas is heated, the atoms it contains move faster and knock together harder. Some of the kinetic energy in one atom may be transferred to an electron in another atom, bumping it up to a higher energy level.

Through photon absorption or collision an electron can jump up one or more energy levels and the atom is in a state of excitation. However, the excited state does not last long and the electron loses energy again by emitting a photon. The bigger the downward jump of an electron, the more energy the photon emitted will carry. Atoms thus return rapidly to their ground state.

Pauli Exclusion Principle

Wolfgang Pauli (1900-1959) from Austria devised the Pauli Exclusion Principle in 1925. It explains how electrons are distributed between electron shells in the atom and their quantum properties.

Each electron has four quantum numbers – three spatial coordinates and a fourth one which is spin or angular momentum. Spin has direction. Pauli's rules state that no two electrons in an atom can have the same values of all four quantum numbers. Electrons fill up electron shells in the atom from the inner shell to higher shells. For two electrons to occupy the inner shell, they must have opposite spin directions with values of positive one-half and negative one-half. An orbital occupied by a pair of electrons of opposite spin is filled and no more electrons may enter it.

Electrons distributed across the energy shells of the atom cannot be squashed into the lowest energy state shell closest to the nucleus. They resist this compression with great pressure. This is what gives matter its rigidity.

The energy states of subatomic particles are forms of motion – the forms of motion include the spin (intrinsic angular momentum) of the particle on its axis, orbital motion around the nucleus and wave motion.

Photons which have no mass can occupy the same quantum state together. Photons can exhibit coordinated group behaviour; for example, a laser beam can consist of photons of a single colour acting together.

2. The electromagnetic spectrum

When photons leave atoms they travel through space as waves but they also behave as particles. Light waves are known as electromagnetic radiation. Electromagnetism is a force, thus photons carry the electromagnetic force. The amount of energy carried by electromagnetic radiation depends on its frequency or wavelength. High energy photons travel at high frequency or short wavelengths.

Different elements or states of the same element absorb and emit radiation in characteristic ways. Spectroscopy is the recognition that each element has its own characteristic series of emission and absorption lines in the electromagnetic spectrum. The emission and absorption of radiation by atoms and molecules is diagnostic of the structure of matter. This allows the matter in our galaxy and other galaxies to be recognized and studied by astronomers.

The electromagnetic spectrum moves from very high energy gamma rays, X-rays and ultraviolet light that may have destructive effects, through visible light in the colours of the rainbow, followed by infrared –the rays of warmth, then radar, microwaves and television waves, and finally radio waves with very long wavelength. This is radiative energy.

Radio waves have the longest wavelengths of metres or kilometres; visible light has wavelengths that are similar to the spacing between atoms (which is why the eye can see them); and X-rays and gamma rays have very short wavelengths that can penetrate atoms and cause damage to them.

Light travels at 300 million metres per second in a vacuum but in a dense material it can be slowed to half this speed. Light waves travel at slower speeds in denser materials such as in water or glass than in air. When light waves cross a boundary between materials the rays bend. This is called refraction. Also, light travels faster in hot air than in cold denser air.

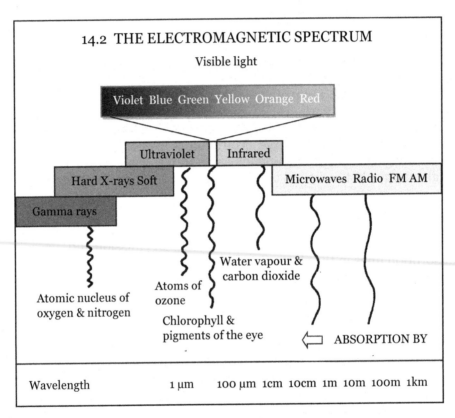

This illustration was adapted from diagrams in Zeilik 2002, pages 101 & 121.

Energy states −nucleons

Just as electrons in an atom can occupy different energy levels, protons and neutrons of the nucleus can also occupy different energy levels. Nuclei have a shell structure like atoms with the protons and neutrons in orbitals. The normal state is the most bound state. The particles of the nucleus may change from a low energy bound state to a higher energy state by input of energy from outside. In a high energy state they may fall back down to a lower level by emitting a specific amount of energy.

The typical excitation energy of the nucleus is a million times higher than atomic excitation energy involving electrons. When the nucleus drops back to its ground state, the quantum of energy emitted is extremely strong. It is called a gamma ray. To induce an excited state of the nucleus, energy must come from a powerful source such as a nuclear reaction.

In the electromagnetic spectrum, visible light, ultraviolet light and X-ray radiation come from electrons, while gamma radiation comes from the nuclei of atoms. Nuclear reactions and radioactive decay produce gamma rays. Gamma rays are extremely strong and harmful to complex molecules, and thus to life.

Energy states –molecules
Molecules also occupy different energy states. A molecule consists of atoms linked together in a particular arrangement by the sharing of electrons (covalent bonds). The molecule can have different energy states depending on how the separate nuclei in the molecule vibrate relative to one another, or rotate bodily. When a molecule changes its vibrational or rotational state, it can emit or absorb a photon. These photons are infrared or radio photons.

Heat
Thermal energy is part of the electromagnetic spectrum. Vibrational motion inside a molecule gives rise to infrared spectra. Random vibrational motion is heat.

Heat may also be produced by the motion of atoms or molecules in a substance relative to each other. The irregular motion of atoms or molecules in a solid substance, liquid or gas produces kinetic energy among the large collection of particles. The more rapid the motion of the particles, the higher the temperature of the substance.

When the temperature of a gas is raised, it undergoes thermal agitation –this means that the speed of atoms in the gas increases and collisions between atoms become more frequent. Collisions excite the atoms, raising their energy state, but they drop down to a lower energy state an instant later with the emission of light. It is for this reason that hot gases shine.

The temperature of a substance indicates the average energy of the particles that make it up. In a gas or liquid, molecules are free to travel in any direction, and often bounce off each other. In a solid, atoms are anchored in a lattice structure. When solids are heated the atoms vibrate while remaining in their positions in the lattice. Thus, collisions between atoms or molecules happen in hot gases, but not in solids.

When a substance is cooled, its atoms move around less. In a gas their speed drops and in a solid their vibrations are reduced. As the temperature drops there is less and less movement until at absolute zero atoms would stop moving altogether. The hypothetical point of absolute zero when energy reaches zero is −273.15 °C or 0 Kelvin.

Radio waves

The rotational motion of nuclei gives rise to the emission of radio photons. In the cold regions of space where molecules exist in interstellar clouds, occasional collisions between molecules cause them to start spinning. These excited molecules emit radio photons that can be observed as characteristic emission lines (Zeilik 2002, page 310). The characteristic radio and infrared radiation emitted by molecules in space allows them to be identified even at intergalactic distances.

X-rays

To complete the picture on the emission of electromagnetic radiation by atoms I will describe the origin of X-rays. When energetic electrons or photons enter an atom and knock out an electron from its orbital in an inner electron shell close to the nucleus, a vacancy is left. The vacancy is filled by an electron from a higher electron shell jumping into the empty orbital. The photon emitted when this occurs is very strong and occupies the X-ray region of the electromagnetic spectrum.

3. Subatomic particles

Subatomic particles are also known as elementary particles. They are studied by particle physicists. Twelve basic subatomic particles have been observed; six are quarks and six are leptons.

Quarks

Protons and neutrons are composed of quarks. There are **Up Quarks** with an electrical charge of $(+ 2/3)$ and **Down Quarks** with an electrical charge of $(- 1/3)$. The proton is made up of two Up Quarks and one Down Quark. This gives the proton an electrical charge of +1: $(+ 2/3) + (+ 2/3) + (- 1/3) = (+1)$.

The neutron is made up of one Up Quark and two Down Quarks. This makes the neutron neutral: $(+ 2/3) + (- 1/3) + (- 1/3) = (0)$.

There are also **Charm** and **Top Quarks** that have the same electrical charge as Up Quarks, but are heavier. There are **Strange** and **Bottom Quarks** that have the same electrical charge as Down Quarks, but are also heavier.

Free quarks do not exist. In nature the quarks making up nucleons in the nucleus are always found in groups of three. In experiments a single quark cannot be isolated. Quarks have a baryon number of 1/3 which means they exist in threes.

Apart from their electrical charge, quarks have a spin of ½ oriented in either of two directions. This will be discussed shortly.

The heavier Charm and Top, Strange and Bottom quarks have been created in particle accelerators and are sometimes produced by cosmic ray collisions in the upper atmosphere; however, they are unstable and quickly decay into Up and Down quarks. Thus, matter that represents most of the mass of the universe is composed of Up and Down quarks.

Leptons

The fundamental particles classified as leptons include the **Electron**. The electron cannot be divided into anything smaller. The electron has an electrical charge of −1.

There are also particles heavier than the electron called **Muon** and **Tau particles**. Muon and Tau are both unstable and have very short split-second lifetimes. The Muon decays immediately into an electron, neutrino and anti-neutrino. The Tau which is heavier than a proton, decays first into a Muon that decays or it decays directly into an electron, tau-neutrino and electron anti-neutrino. The decay of Muons and Tau occurs via the weak force.

Electrons are very light particles compared to the other particles mentioned so far. When atoms are raised to high energy states and ionized, electrons are freed from the outer orbitals of the atom and can move independently through space. Electric currents pass through wires by freed electrons transferring current. Lightning observed in the atmosphere is caused by the passage of freed electrons. Synchrotron radiation is a stream of high-speed electrons moving through space.

Neutrinos
A lepton with no electrical charge is the **neutrino**. Neutrinos are extremely small, but they can carry large amounts of energy. The neutrino may have no mass like photons or have a very small mass. There are three types of neutrino: the **Electron Neutrino, Muon Neutrino** and **Tau Neutrino**.

When a proton combines with an electron to become a neutron, it results in the release of one neutrino. This process occurs when the core of a star collapses to become a neutron star during a supernova explosion. During supernova explosions space is flooded with neutrinos. Neutrinos pass straight through nearly all the objects they encounter because they are so small and have no electrical charge.

Neutrinos are also released during nuclear reactions. The conversion of hydrogen into helium in the core of stars such as the Sun results in the emission of neutrinos[6].

_____

[6] There is the problem of the missing neutrinos —experiments on Earth only detect one third to two thirds the number of electron-neutrinos expected to arrive from the Sun. However, on a recent programme I saw, it was stated that Snowlab in the Sudbury Mine, Canada has detected the missing third of neutrinos from stars.

When elements undergo radioactive decay called beta decay, neutrons in the nucleus are converted into protons with the release of electrons and **anti-neutrinos**. Each neutron that becomes a proton emits one electron and one anti-neutrino.

Neutrinos and anti-neutrinos have a ½ unit of spin, no electrical charge, no or very little mass and they travel at the speed of light. Neutrinos have left-handed spin only, while antineutrinos have right-handed spin only.

Electrons are paired with neutrinos. Often the term 'neutrino' is used to refer to both the neutrino and its antiparticle.

Positrons

The positron has the same mass and the same spin as the electron, but the opposite electrical charge. It is an antiparticle. Positrons do occur naturally unlike most other forms of antimatter. When cosmic rays collide with atomic nuclei in the upper atmosphere, the release of energy creates showers of particles that include positrons and electrons. When a positively charged positron comes close to a negatively charged electron they annihilate and the energy of the pair is converted into a gamma ray photon.

Magnetism and spin

Electrons possess an internal motion (or angular momentum) called spin. The spin may be $-$ ½ or $+$ ½. Thus, electrons behave like spinning electrical charges each with its own magnetic moment. The spin of the electron provides it with a directional orientation. The electron has a magnetic moment along its spin axis.

Electrons are subject to the electromagnetic force, the force of gravity and the weak force. Each electron orients itself either parallel or anti-parallel to the magnetic field of the rest of the atom.

Protons and neutrons also have spin with quantum numbers of ½ and two spin orientations. They also have a magnetic moment associated with their spin.

Quarks have spins of ½. To produce a proton or neutron with a spin of ½, the quarks composing the nucleon must align such that two of the three spins cancel each other out leaving a net value of ½.

The overall spin of a nucleus is zero if the nucleus has an even number of protons and neutrons. Nuclei with an odd number of nucleons always have a half-integer spin. A nucleus with a spin also has a magnetic moment aligned along the spin axis.

Magnetic field lines are always closed loops running from the north pole to the south pole of the magnet. There are no (known) magnetic monopoles – if you chop a bar magnet in half, you obtain a north and south pole for each half. Grand-scale magnetism results when the magnetic fields of single atoms are all aligned – this is produced by the spin axis of electrons. A changing electric current produces a magnetic field, while a changing magnetic field produces an electric current. The links between electricity and magnetism were described in James Clerk Maxwell's four equations on electromagnetism.

The magnetic field produced by the nucleus is not as strong as the magnetic field produced by the electron; however, it produces observable effects. The technique of magnetic resonance is based on the fact that in a strong magnetic field nuclear spins tend to align, and it uses the hydrogen nucleus because its one proton gives it the largest magnetic moment for its spin.

Unstable particles
Particle accelerators such as those at CERN (European Organization for Nuclear Research) near Geneva convert energy into massive particles. About 200 subatomic particles have been found, however, they are unstable and only exist for split-seconds. These unstable particles decay rapidly into protons, neutrons and electrons plus neutrinos.

An example of an unstable particle is the meson. A meson is composed of a quark and an antiquark. Mesons are produced in cosmic ray interactions and in high-energy particle accelerators. All mesons are unstable with lifetimes ranging from 10^{-8} to less than 10^{-22} second and they vary widely in mass. The pi-meson or pion decays into two photons.

A neutron outside an atomic nucleus decays within a few minutes ejecting an electron and an anti-neutrino to become a proton. Only protons are stable; no one has ever seen a proton decay. Protons and neutrons are made from Up and Down quarks since only these two are stable; the other four quarks are unstable.

This leaves a material world essentially made up of Up and Down Quarks, electrons, electron neutrinos and antineutrinos. Therefore, out of the 200 subatomic particles known and 12 main subatomic particles, only five are stable and are found as the constituents of matter.

Terminology

Protons, neutrons and other massive particles composed of quarks are referred to collectively as **hadrons**. Hadrons include **baryons** composed of three quarks and **mesons** composed of a quark and an anti-quark. There are also **bosons** that are massless particles that carry forces.

In Supersymmetry Theory particles with ½ spin that include quarks and leptons are called **fermions**. Bosons are particles with 1 to 2 spin, and the Higgs particle is postulated as having no spin.

4. Forces

Any physics book will tell you that there are four forces in the universe: the electromagnetic force, the strong nuclear force, the weak nuclear force and gravity.

Forces are carried by massless particles called **bosons**. The only well-documented and well-known boson is the photon. Whereas the fundamental particles have half spins, the bosons have whole number spins.

The electromagnetic force

The photon is a boson that carries the electromagnetic force between electrically charged particles. It has a spin of 1. Photons carry both energy and momentum. The electromagnetic force binds negatively charged electrons to positively charged atomic nuclei. The electromagnetic force gives rise to the bonding between atoms (or charged ions) to form molecules. Matter in the universe interacts via the electromagnetic force.

The strong nuclear force

The strong nuclear force binds quarks together within protons and neutrons. The strong nuclear force also keeps protons and neutrons together within atomic nuclei despite the mutual repulsion of the positively charged protons. The strong force does not affect electrons or neutrinos since its effects are only felt within the nucleus of the atom.

When quarks are close together they can move freely, but as soon as they move apart the strong force exerts itself to hold them within the proton or neutron. When protons are brought close together, they initially repel each other due to their like charges, but if they are pushed together closer, the strong nuclear force locks them together. Thus, at close distance within the nucleus the strong force is strongly attractive and binds the nucleons together as a nucleus. However, if the protons are pushed even closer together, they resist and act as hard spheres. Thus, the strong nuclear force is very weak outside the nucleus, very strong at close range within the nucleus, but it has a fixed limit as to how close it will allow protons to go.

Because of the strong nuclear force, nuclear binding energies are a million times stronger than the binding energies of electrons in atoms. For this reason the amounts of energy that can be released in the transformation of the nucleus are much larger than the chemical energies released by the transformation of electron patterns in atoms.

The boson thought to be responsible for the strong nuclear force is called the gluon. It has not yet been directly observed.

The weak nuclear force

Since the 1930s particle physicists have attributed a type of radioactivity known as beta decay to a force acting within the nucleus named the weak nuclear force. Beta decay involves the transformation of a neutron into a proton with the concomitant release of an electron and an anti-neutrino.

Free neutrons are unstable and have a mean life of 896 seconds before decaying into protons, whereas neutrons bound in atomic nuclei can be stable and have lifetimes of millions of years, although they eventually undergo beta decay. Highly radioactive elements have neutrons with lifetimes of a thousandth of a second. Thus, radioactivity in elements is thought to occur by the weak force.

In stars such as the Sun where the temperature and density of matter are high, the nuclear fusion process going on also involves the weak force. Two protons interact via the weak force to form a deuterium nucleus, which reacts further to generate helium.

The weak force is effective only within the nucleus and in radioactive decays its strength is about 100 000 times less than the strength of the electromagnetic force. Thinking on the weak force has recently changed with the theory that the electromagnetic force can be combined with the weak force as the Electroweak Force. This theory will be explained in the next section.

Gravity

Gravity is the attraction of matter to other matter such that a body will draw another body to itself. Gravity compared to other forces is very weak. In terms of the mass and electrical charge of a proton, the strength of gravity is 10^{-40} (this is the gravitational fine structure constant). [7]

The General Theory of Relativity proposed by Albert Einstein (1879-1955) is a theory of gravitation. In General Relativity gravity is the

---

[7] Mass is a measure of how much matter an object contains. The mass of an object is the sum of the masses of all the atoms in it. Weight is a measure of the force of gravity pulling a mass down. The weight of an object on the Moon is less than the weight of the same object on Earth because the gravitational force exerted by the Moon is less; however, the mass is the same.

distortion of spacetime – gravity is the curvature of spacetime in the presence of mass. Einstein's theory is about the structure of the universe. In a nutshell, "space tells matter how to move; matter tells space how to curve," (Gribbin & Rees 1991, page 20).

In the Theory of General Relativity the dimensions of space and time are combined into a spacetime grid. The speed of light is fixed and nothing can exceed it. When moving and accelerating, it is spacetime that distorts to maintain the fixed speed of light.

One of the predictions of General Relativity is that light can be bent by a gravitational field. This has been found to be so.

Gravity affects the large-scale structure of the universe, but its weakness compared to the other forces casts doubt on whether it really is a force. Einstein did not present gravity as a force. If gravity is a force, the boson which mediates it –the graviton has not yet been found.

5. Hypothetical particles

According to Quantum Field Theory all fields have particles associated with them. The strong force and the weak force are presently understood by positing the existence of hypothetical particles. These hypothetical particles allow the mathematical equations to balance, but they have never been observed in experiments. Gravity also has its particle, the graviton; however, the adding in of gravity to equations tends to cause them not to work anymore.

- The strong nuclear force is thought to be associated with a boson called a gluon. The action of gluons is described by the Theory of Quantum Chromodynamics (QCD).

- The weak nuclear force (responsible for beta decay) is thought to be mediated by W and Z particles. These are virtual particles.

- The force of gravity that affects all particles and matter with mass is thought to be mediated by the graviton.

Quantum Chromodynamics Theory (QCD)

Quantum Chromodynamics is an analogy of Quantum Electrodynamics (QED) formulated in the 1970s. Quantum Chromodynamics predicts that the interaction of quarks occurs by the transmission of massless force-carrier particles called gluons. Gluons are said to carry a strong charge called colour. The colours are red, green and blue and anti-colours. This is an analogy to the primary colours of light, although Quantum Chromodynamic colour has nothing to do with colour in the real sense. It is difficult to say where this analogy is going.

Electroweak Theory

The Electroweak Theory unites electromagnetism with the weak force as manifestations of the same thing. In this theory the weak force is mediated by positively charged and negatively charged W particles and neutral Z particles. The W and Z bosons would be massive particles, thought to be a hundred times larger than protons. The W and Z particles are virtual particles –this means that they decay immediately into an electron and anti-neutrino in a lapse of time too brief to measure. The W and Z particles are theoretical and inferred, while the electron and anti-neutrino are observable.

CERN is looking for W$^+$, W$^-$ and Z^0 particles. So far it has only found their products (electrons and anti-neutrinos) because W and Z are thought to be so transient.

Matter in the real world appears to be asymmetrical; however, Electroweak Theory proclaims that there is symmetry, although it may be a hidden symmetry. Symmetry between particles and antiparticles is one of the main concepts in theories of particle physics today.

The Higgs boson

Current theories in particle physics build in more symmetry than is actually observed in the particle world. These symmetries are said to exist at very high energy levels that no longer exist today; for example, in the first instants of the Big Bang. The breaking of such symmetry requires the presence of extra particles in the universe (Gribbin & Rees 1991, page 114). These particles have not yet been found.

At CERN there are currently hopes that the new super-collider (Large Hadron Collider) will be able to generate such high energy collisions between particles that the hypothetical bosons will be found. The search is on for the Higgs boson. The Higgs particle is necessary to the Electroweak Theory because it allows the symmetry of the electroweak force to be broken. It would also have broken the hypothetical symmetry between matter and antimatter at the Big Bang.

The hypothetical Higgs particle is postulated to carry the Higgs field, a theoretical field that permeates space and endows all subatomic particles with mass. It is hypothesized that elementary subatomic particles have different masses due to the different strengths of interaction they have with the Higgs field.

It is thought that the Higgs boson must be very massive – its mass must be greater than the energies already reached experimentally in particle accelerators or it would already have been found, it is claimed. The mass of the Higgs boson is not thought to exceed the equivalent of a trillion electron volts.

There have been tentative claims in the last years that evidence shows that it might have been found – although no one is entirely sure.

6. Theories in particle physics

Current theories in particle physics are generally based on the idea that there is matter and antimatter. Like mirror images, a particle and its antiparticle would be related by various kinds of symmetry. If a particle is positively charged, its antiparticle is negatively charged or vice versa; and if a particle is spinning clockwise, its antiparticle is spinning anticlockwise or vice versa. Symmetry in time would mean that mathematically because of their negative energy, antiparticles would be moving backwards in time from the future to the past. These antiparticles exist in mathematical equations.

A real antiproton was created at CERN and Brookhaven Laboratory in America in 1995. The creation of antimatter has continued at CERN in Switzerland and Fermilab near Chicago. When a beam of particles meets a beam of antiparticles, they annihilate each other in a flash of pure energy.

It is of no real surprise that the universe is formed from matter with no detectable admixture of antimatter, since if this were not so, the universe would have ceased to exist in a puff of energy a very long time ago, and we would not be discussing it now.

Physicists have proposed that at the beginning of the universe, there was matter and antimatter in equal proportion arising from (nothing?). Something happened to break the symmetry between matter and antimatter such that there occurred an excess of matter, while all the other matter and antimatter annihilated itself. This excess matter, even if quite small, became our universe.

Various theories are currently in fashion:

The Standard Model
The Standard Model of Particle Physics is a theory of elementary subatomic particles and fundamental forces with the interaction occurring between elementary particles through the exchange of massless particles.

The Standard Model is the combination of the Electroweak Theory and the Quantum Chromodynamics Theory of the strong nuclear force. Both these theories are gauge field theories which means that they describe the interactions between particles in terms of the exchange of intermediary 'messenger' particles. The Higgs particle is a major ingredient of the Standard Model. The Standard Model does not include gravity.

The Standard Model is not a complete theory. It has many ad hoc explanations. The hope of physicists is to find a grand unified theory which will explain the strong, weak and electromagnetic forces under a single theoretical framework.

Supersymmetry
An alternative to the Standard Model is Supersymmetry. In Supersymmetry Theory quarks and leptons are called fermions with ½ spin, bosons of gauge fields have 1-2 spin and the Higgs particle has 0 spin. The theory allows all the forces to be united, but it requires supersymmetric counterparts to all the known particles doubling the

number of particles. The particles added are the squark, slepton, photino, wino, zino, gluino and gravitino. These particles must all be heavy to have escaped detection.

GUTs

Grand Unified Theories or GUTs seek to unify the strong, weak and electromagnetic forces. GUTs do not include gravity. GUTs are based on mathematical symmetries, and they introduce the X boson.

The **X boson** is a hypothetical particle that would have broken the symmetry between matter and antimatter at the beginning of the universe. The X boson would be a massive particle that decays in a slightly imbalanced way to give a slight overproduction of matter. It is proposed that X bosons may also interact with protons and cause them to decay.

The strengths of the forces do not converge exactly, but it is postulated that new effects may come into play causing the forces to converge at high energies. However, the energies involved would be more than a million million times what can be produced experimentally now.

Grand Unified Theories describe the interactions of quarks and leptons. It is proposed that quarks can decay into leptons and that the proton can decay. Experiments designed to test this hypothesis have not found any evidence of protons decaying.

TOE

TOE is the **Theory of Everything.** It unites Supersymmetry and **String Theory**. **Superstring Theory** is a TOE. String Theory proposes that subatomic particles can be represented as vibrations in one-dimensional strings. Different String Theories have different numbers of dimensions, but a String Theory may have up to 11 dimensions instead of the usual three known spatial dimensions of our universe. This allows strange things to happen.

The variety of fundamental particles arises because of the pattern of vibration of strings which may be open-ended or in closed loops. The strings vibrate in many different directions due to the extra dimensions available.

String theories incorporate gravity as well as the other three forces. Superstring Theory is, however, only a mathematical construct based on no experimental evidence. **M-theory** incorporates in addition to strings, membrane-like objects of various dimensions called branes. String Theory and M-theory have no falsifiable hypotheses.

Comment

The attraction of these theories is to arrive at a unified theory of particles and forces such that the whole universe could be described by one equation. This one equation would eliminate the 19 or so constants of nature which are specific numbers. The speed of light, the mass of the proton, the electrical charge of the proton and the gravitational fine structure constant are examples of constants of nature.

Theories in particle physics have given rise to large numbers of hypothetical particles. Photinos, gravitinos and axions have been described by Gribbin & Rees (1991, page 117) as "exotic inhabitants of the particle zoo". They have not yet been detected.

One could ask the question, why is there so much enthusiasm for so many hypothetical particles?

Hypothetical particles make the Standard Big Bang Model equations work; they are useful for balancing equations. Apart from this, the answer is that they play another intellectual role; they lend support to theories concerning dark matter in the universe. Axions, neutralinos, WIMPs (Weakly Interacting Massive Particles) and other hypothetical particles are all candidates for dark matter. Dark matter is important to cosmologists. (Dark matter will be discussed in Chapter 16).

Cosmological philosophy

In addition to the theories about the universe listed above, the Multiverse is an alternative explanation in vogue amongst some cosmologists (see footnote[8]). The Multiverse invokes the arising of a sequence of multiple universes or the existence of parallel universes as a way of explaining away the fundamental constants of nature.

---

[8] Carr, B. (ed) (2007) *Universe or Multiverse?* Cambridge University Press

Hugh Everett's Many Worlds Interpretation of Quantum Mechanics formulated in 1957 has an infinity of parallel universes in which everything which can exist, simultaneously exists. M-theory as a multi-dimensional extension of String Theory has tiny universes popping out of different dimensions.

One version of the Anthropic Principle argues that if there are many or even an infinity of universes each with different physical laws, at least one or some of these universes must have physical constants that have allowed them to develop life – our universe is one of these.

The Multiverse is a conglomeration of philosophies about the nature of reality discussed in the place of scientific theories.

7. Quantum Mechanics

Quantum Mechanics describes the behaviour of matter and radiation at the atomic scale.

The application of Quantum Theory to the interaction between electrons and radiation required a quantum treatment of Maxwell's field equations (of 1873) which describe electromagnetism and the Relativistic Theory of the electron formulated by Paul Dirac in 1926. The resulting Quantum Field Theory was Quantum Electrodynamics (QED).

In 1926 the Austrian physicist Erwin Schrödinger formulated an equation for wave mechanics. Schrödinger's theory tells how to write down an equation for the wave function of any physical system in terms of the masses and charges of its components. Using the wave function, one may compute the energy levels and other observable properties of the system. Schrödinger's equation for wave mechanics along with work by German physicists Werner Heisenberg and Max Born launched the general theory of Quantum Mechanics.

Electromagnetic radiation exhibits wave properties such as interference and diffraction, as well as particle properties in that the particle such as the photon can be pin-pointed in space. A famous experiment performed by Young involving a beam of light and two slits demonstrates both the wave and particle properties of light. The experiment can also be conducted with a beam of electrons with the same results.

This experiment shows that light or photons have particle-like properties, while matter such as electrons have wave-like properties. The idea that matter could behave as waves was suggested by Louis-Victor de Broglie in 1924. These insights are fundamental to Quantum Mechanics.

The strange thing about the experiment to demonstrate particle-wave duality is that when a single electron is passed through the slits and located as a particle passing through one slit, the wave function collapses, while when it is not measured the interference pattern of waves is maintained. If a single electron must only pass through one of the two slits, it is not known how an interference pattern could be produced, or why this interference pattern should disappear when it is observed which slit the electron passed through.

Another feature of Quantum Mechanics is that it is impossible to measure a system without disturbing it. When investigating the subatomic components of the atom it was found that the exact position and energy of a subatomic particle could not be determined exactly, it could only be estimated as a probability. This situation was realized in 1927 by Werner Heisenberg and he named it the Uncertainty Principle.

Heisenberg's Uncertainty Principle

The Uncertainty Principle states that you cannot simultaneously measure both the position and momentum of a subatomic particle. In other words, you cannot measure the particle's energy at an accurate time. If you know the position of the particle, then the momentum (or speed) is uncertain and if you know the momentum, then the position is uncertain. Both can be measured within certain bounds, but as probabilities.

The reason for this is that in order to measure the motion of a subatomic particle such as a neutron it must be tracked by radar. Radar tracks the particle by bouncing electromagnetic waves off it. Gamma rays will locate the particle very precisely because gamma rays have short wavelength. However, the gamma rays hitting the neutron will act like photon bullets. The high energy of gamma ray photons will accelerate the particle unpredictably. Thus, you may have determined the position of the particle when the gamma photon hit it, but its speed will have changed due to the process of observation (Baker 2007, page 105).

You could use lower energy photons to locate the subatomic particle; however, longer wavelength will mean that the position of the particle is not fully known, even if the acceleration of the subatomic particle was minimized. Therefore, how ever one adjusts the experiment, the particle's position and speed cannot be known simultaneously.

In order to observe something physicists must bounce photons off it. In the subatomic world, this transfer of energy involved in the act of observation disturbs the particle's original motion.

The Copenhagen Interpretation of Quantum Mechanics

The Uncertainty Principle of Heisenberg was expanded by Niels Bohr (1885-1962) from Denmark into what has become known as the Copenhagen Interpretation of Quantum Mechanics. The Copenhagen Interpretation is based on the view that the indeterminacy observed is fundamental to matter at the quantum level and is not the result of inadequacy in the measuring apparatus or in scientific knowledge. The proponents of this view believe that the electron in the atom has no definite position or velocity, and therefore that matter is ultimately not fully determined.

The Copenhagen Interpretation has become the orthodox view of Quantum Mechanics today.

In the philosophy of Bohr, reality is determined by its measurement. Bohr's belief was that quantum systems exist in probabilistic limbo until observed, at which point they adopt their final state. Baker (2007, page 116) explains, "Before being filtered by observation, the system exists in a combination of all possible states of being."

The Copenhagen Interpretation was strongly opposed by Erwin Schrödinger and Albert Einstein in the 1930s who set about showing how absurd the Copenhagen Interpretation is with thought experiments such as 'Schrödinger's cat'.

144

Schrödinger, Einstein and others rejected the notion of intrinsic uncertainty, and held the more boring belief that the physical world has real properties whether one measures them or not. I agree with the boring interpretation of Quantum Mechanics.

Heisenberg thought that the uncertainty of a quantum system arose from the mechanics of the measurement act itself. I am not sure of the wisdom he had in objectifying which led to reifying a negative experimental result by giving it the name 'Principle'.

In order to observe the position and momentum of a subatomic particle within an atom, one must employ photons. When the photons achieve their task by hitting the subatomic particle, they have the effect of energizing the particle unpredictably. It may be that the experiment involving two slits and beams of photons or electrons has a similar explanation. The position of the single electron passing through one of the two slits can only be determined by use of a detector. However, the detector itself may deflect the path of the electron or change its spin axis such that the wave pattern collapses and interference is no longer produced.

Conclusion

I have attempted in this chapter to give a concise account of the structure of matter by describing what is known to exist through observation and experiment in particle physics. The chapter started with a description of the atom, electromagnetic radiation and the subatomic particles of which the atom is composed.

The four fundamental forces of nature were described: the electromagnetic force, strong nuclear force, weak nuclear force and gravity. New theory combines the weak nuclear force with the electromagnetic force as a single force named the Electroweak Force mediated by virtual particles. It is not really known whether gravity is a force or not since General Relativity does not present it as a force. General Relativity presents gravity as a distortion of spacetime. There may, in fact, be only two fundamental forces.

I suggest that there may be only two fundamental forces – the electromagnetic force which acts upon the electron orbital structure of the atom and on the chemical interaction of atoms; and the strong nuclear force which acts on the nucleus of the atom holding quarks together within protons and neutrons, and these nucleons together within the nucleus.

It may be that the beta decay of neutrons into protons with the concomitant release of electrons and antineutrinos could be likened to the Second Law of Thermodynamics which involves a running-down of the system; a move towards disorganization. Beta decay may not necessitate any input of energy or any force for it to take place. It would simply be the random decay of elements back into the original types of atom i.e. an element with a lesser number of neutrons than the radioactive element produced from it.

Beta decay returns elements back to their original simpler state. Likewise, gravity acting upon mass to bring bodies together – squeezing out space – with the ultimate consequence of gravity crushing atoms within black holes and squeezing out the space within the atoms – would simply be a return to the original state of the universe. The universe (according to the hypothesis that I will present in Chapter 15) started out as a very dense localized area containing almost all of the mass in the universe. With the input of energy, space and time were created from this initial dense state.

Quantum Mechanics has become a strange mixture of science and philosophy, observation and belief. Einstein was disturbed by the apparent denial of the objective reality of the atomic world encapsulated in the belief of followers of the Copenhagen Interpretation of Quantum Mechanics. The Copenhagen Interpretation asserts that somehow the electron's position or velocity only comes into existence when it is measured. Einstein voiced his objection to intrinsic randomness in his famous statement that God "does not play dice." Einstein also voiced his belief in an exactly engineered universe in the statement "God is subtle but he is not malicious."

(See footnote[9] on Einstein's personal beliefs).

---

[9] Albert Einstein (1879-1955) was Jewish in religious and cultural background. He did not practice Orthodox Judaism. Einstein was a Pacifist in Germany before World War II, anti-Nazi and a Zionist. Einstein promoted Pacifism, although he advocated nuclear armament of the Allies against the Nazis to get the atomic bomb before the Germans managed to develop it.

To my mind it is clear that for God, knowledge of the location and momentum of a subatomic particle would not be uncertain because God does not need to use gamma rays in order to observe them. God sees, but does not have physical eyes. Both Christians and the majority of believers in other religions hold that God knows His creation. He is omniscient. We humans must do experiments in order to observe and know. At the subatomic level the limits to our knowledge are reached due to the workings of physics. But what for us is uncertain is not uncertain for God.

The Uncertainty Principle and Copenhagen Interpretation of Quantum Mechanics have led some Theists to propose that God acts in the world by means of Quantum Uncertainty. In this view, uncertainty at subatomic level would allow God to 'flick the switch' and determine outcomes without it being perceived as unnatural.

In my opinion God does not require things to be uncertain or undefined at the subatomic level to be able to act. However, I do believe that God acts within the universe by interaction at the subatomic level – ordering atoms and molecules. Therefore, the subatomic level could be seen as providing the seam between nature and divine action.

I believe that God created matter out of nothing as subatomic particles which then formed atoms. Each subatomic particle was created with its own specific properties in terms of mass, electrical charge and spin. The subatomic particles of which the universe is composed are mainly Up Quarks and Down Quarks forming protons, electrons, neutrinos and antineutrinos. All other subatomic particles are unstable.

The atom is composed mainly of empty space; however, this empty space contains motion – all subatomic particles have various types of angular momentum. Matter is an interaction of motions. Atoms interact via the emission and absorption of photons of light which are quantized energy packages travelling as waves.

Einstein denied atheism and expressed a belief in God as the harmony of what exists. He passionately believed in scientific truth and dedicated most of his life to theoretical physics. Einstein gave many lectures worldwide on science and on Pacifist politics. The balance of his life was summed up when he said at the end of his life "Politics are for the moment. An equation is for eternity."

In the Theory of Nanocreation, matter was created asymmetrical. Asymmetry is found in the organic carbon molecules of living cells: amino acids are exclusively L-form (left-handed) and polysaccharides are exclusively D-form (right-handed). The opposite – D-form amino acids and L-form polysaccharides can be made in test-tubes, but these carbon molecules do not occur naturally in living things. (The subject of chirality in organic molecules was discussed in Chapter 11, section 14). I believe that God created life with asymmetrical carbon molecules. In the same way the universe is composed exclusively of matter – positively charged protons and negatively charged electrons; it is asymmetrical. I believe that matter was created that way, although antimatter – negatively charged protons and positively charged electrons can be made in particle accelerators.

Thus, in the Theory of Nanocreation matter was created out of nothing, but there was never a phase when it disappeared into nothing through annihilation with antimatter. The Theory of Nanocreation does not proclaim that symmetry (between matter and antimatter) ever existed, and it does not require exotic hypothetical particles to break symmetry. The hypothetical Higgs boson or X boson are not necessary, and Strings are not required.

The observed universe today is asymmetrical and its functioning is dependent upon asymmetry; I believe that it was designed this way.

Bibliography
Baker, Joanne (2007) *50 Physics Ideas You Really Need to Know* Quercus

Encyclopedia Britannica 2011 Standard Edition: Anti-neutrino; Atom; Dirac P.A.M; Einstein, Albert; Gluon; Higgs particle; Mechanics; Meson; Pauli Exclusion Principle; Physical science: Quantum mechanics; Quantum Chromodynamics (QCD); Quantum field theory; Quantum mechanics; Radiation; Standard Model; Subatomic particle; String theory; Unified field theory; Weak force.

Gribbin, John & Martin Rees (1991) *Cosmic Coincidences: Dark Matter, Mankind and Anthropic Cosmology* Black Swan

Zeilik, Michael (2002) *Astronomy: The Evolving Universe* 9th edition Cambridge University Press

CHAPTER 15

THE EXPANDING UNIVERSE

Introduction

There was a timeless moment before the beginning when the dimensions of space had not yet opened up and time had not yet begun. It was a moment of eternity when God alone existed because He is existence. I AM. I am Being, and Being has no need of any other justification than Himself.

Out of the void and the nothingness, God created matter.

God created subatomic particles and the void was not a void. The world was in deep darkness, and into this darkness came light. Light and heat energy flooded the universe and motion created space.

God created subatomic particles as different kinds; each with their own intrinsic motions which gave them different properties. He created positively charged particles and negatively charged particles. God set the quantity of matter and the mode of interactions between matter for the unfurling of future events. God created the atom; when the last ingredient was added it went BANG and this was the beginning. God saw that His creation was very good.

1. Cosmic expansion

The universe is expanding. We know this because telescopes show that galaxies are receding away from us with a speed of about one-fifth the speed of light. Galaxies in the further limits of the universe are receding almost at the speed of light. As galaxies recede away from each other three-dimensional space is opening up. Expansion of the universe creates space and time.

The rate of expansion of the universe is generally estimated by astronomers to be 20 km per second per million light-years. This means that the expansion increases by 20 km per second for every million light-years in distance (Zeilik 2002, page 142).

The implication of expansion is that the universe had a beginning at a specific moment in the past. This notion contrasts with the commonly held belief that prevailed until the end of the 1950s that the universe is infinite both in extent and duration. Using the rate of expansion given above, and assuming that the rate of expansion has been constant, all the matter of the universe now found spread out in galaxies would have been coalesced into one place 15 thousand million years ago. Fred Hoyle, who promoted the Steady State Theory of the Universe (see footnote[10]) and believed the universe to be infinite in time and space, called this point the big bang in utter scorn of those who, following Georges Lemaître (1927) believed in it. His opponents did not take exception to the insult and they liked the name so they kept it.

The age of the universe is estimated to be between 10 and 20 thousand million years old or 10 to 20 American billion years old [11]. The size of the universe is the age of the universe multiplied by the speed of light (Cosmos page 288). So, the present idea is that the universe is vast, but not infinite.

The universe is finite but unbounded if there is sufficient matter in the universe to curve spacetime back onto itself. The condition of unbounded means that when light travels outwards and reaches what would be the 'edge of the universe', because light travels on a curved path in the presence of matter, the light will continue its curved path indefinitely and so never come to an edge. For this reason, there is, in fact, probably no edge.

---

[10] The Steady State Theory was first proposed by Sir James Jeans in about 1920, then revised in 1948 by Hermann Bondi and Thomas Gold with contributions from Fred Hoyle. It was based on the 'Perfect Cosmological Principle'. This Principle states that the universe is homogeneous in both its spatial and temporal dimensions. This Principle is at odds with the Second Law of Thermodynamics andthe predicted Heat Death of the universe. The Steady State Theory took care of increasing entropy in the universe with the hypothesis that matter is continually being created in the form of new stars and galaxies. The creation of matter keeping apace with the expansion of the universe would keep the average density of matter in the universe constant and counteract the Heat Death. This theory has now been totally dropped.
[11] An American billion is a thousand million, while an English billion is a million million. I am pointing this out so as not to get confused with the number of zeros. For example, English radio astronomer Sir Bernard Lovell writing in 1959 took a billion to be a million million. I heard on the radio that someone working in a bank once got confused on this point and nearly paid out far too much money!

150

Time is a created entity. Time does not exist without motion. Time in the universe is counted by the expansion rate of the universe. Local Earth time is counted by orbits of the Earth around the Sun (years) and rotation of the Earth on its axis (days). Time is a beginning point to now.

Spacetime has particular features. Spacetime exists in a finite universe. The infinite is timeless. The universe cannot be infinite in the sense of starting at a certain point and 'going on forever'. The reason for this is that the universe is cooling down and when it is completely cold all motion will stop. When all motion stops, time will stop. This is the cosmic heat death predicted by the Laws of Thermodynamics.

A peculiar feature of spacetime is that looking out into space at distant galaxies with the aid of a telescope is looking back in time. Galaxies observed far away are seen as they were a long time ago. The outermost objects observed exist in the early universe, and no longer exist now. This is because light set out from these galaxies and objects thousands of millions of years ago, and travelling at a finite speed (the speed of light), the light is only just reaching us now. Astronomers are blessed with the ability to look into and see the past happening now!

2. The Big Bang Model

The General Theory of Relativity proposed by Albert Einstein in 1916 indicated that the universe should be expanding. This was pointed out by the mathematician and cosmologist Alexander Friedmann.

Edwin Hubble working at the Mount Wilson Observatory in California was the first person to discover the existence of other galaxies outside the Milky Way – our own galaxy, and in 1927 found that all of these galaxies are receding away from our galaxy (except for some local galaxies which are moving towards us). The galaxies were also moving away from each other. This was observable proof of the expansion of the universe.

151

In the same year, 1927, Georges Lemaître (1894-1966) a Belgian astronomer and ordained Catholic priest proposed that the universe had expanded from what he termed the primeval atom [12].

George Gamow (1904-1968) a Russian-born American reworked the findings of Friedmann, Hubble and Lemaître with his knowledge of nuclear physics to formulate his theory of a primeval thermonuclear explosion for which he used the at-the-time derisory term, Big Bang. Gamow believed that this explosion gave rise to the main elements found in the universe. The paper containing this theory was published in 1948 [13].

Stephen Hawking took up the Big Bang theme in 1965 after reading Roger Penrose's theorem on black holes. Penrose had showed that theoretically a collapsing star shrinks down such that all the matter of the star is compressed into a region of zero volume. Hawking called the end of spacetime in a black hole of this sort, a Singularity. Hawking realized that if the direction of time was reversed such that collapse became expansion, then the expanding universe must also have begun in a Singularity.

According to the Singularity argument the star matter contained in a black hole will keep on collapsing gravitationally until it has no volume. Although it has no volume, this matter retains the original mass; consequently its density becomes infinite. The laws of nature break down at infinity and this point is called a Singularity. A Singularity is beyond natural law. Applied to the beginning of the universe, the Big Bang is envisioned as having started as a single point and yet this point contained everything that became the universe. In the 1970s it became generally accepted that the universe had not only begun in a large explosion, but also from a Big Bang Singularity.

---

[12] Books by Georges Lemaître include *Discussion sur l'évolution de l'univers* (1933); *L'Hypothèse de l'atome primitif* (1946).
[13] Gamow, G & R. Alpher (1948) The Origin of Chemical Elements. *The Physical Review* 1st April 1948.

Phases in the Big Bang

The presently accepted Standard Big Bang Model presents the universe as having started out as a fire ball consisting of pure energy. Time is taken back to 10^{-43} seconds, beyond which calculations cannot be made. The temperature near the beginning was 10^{12} K (1 000 000 000 000 or one million million degrees). This immensely hot, dense locality exploded. As the photons expanded outwards the temperature dropped and this caused phases to occur with energy condensing into matter and anti-matter.

The Standard Big Bang Model purports to explain the origin of the subatomic particles of which the universe is composed, as follows:

At first the energy of the photons was so great that they were able to produce protons by collision. The requirement for this to happen is that the energy of the colliding photons equals or exceeds the mass of the particles produced. During this Heavy-Particle Era pairs of particles and antiparticles annihilate each other. The universe continues to expand and cool.

In the next phase the remaining photons are converted into electrons which since they have 1800 times less mass than protons require less energy for their fabrication. During this Light-Particle Era electron-positron pairs are made. The protons and electrons interact to generate neutrons, but the neutrons also decay back into protons and electrons.

In the third phase neutrons and protons react to form deuterium (^{2}H). Further reactions convert deuterium into helium resulting in 25 % helium by mass compared to all forms of matter. Some beryllium and lithium are also formed, and some deuterium is left over. After a few minutes this primordial nucleosynthesis ends because the temperature has dropped too low for fusion to occur.

Matter in the universe is ionized at first due to high temperature. But after a million years the radiation temperature has dropped to 3000 K (Zeilik 2002, page 473) at which point nuclei begin to capture electrons to form neutral atoms. This process is called recombination. When neutral hydrogen atoms formed, radiation became dissociated from matter. At this point the radiation continued its journey through space and its remnant is known as the cosmic background radiation. This ended the Radiation Era. (Some accounts place the formation of

153

atoms at 300 000 or 400 000 years after the beginning, rather than a million years).

The final era is the Matter Era when matter became clumped into clouds that became galaxies. In this era gravitation came into play. The clouds of matter are thought to have undergone gravitational collapse to form galaxies. Further gravitational collapse formed stars. The first star is thought to have formed one thousand million years after the beginning.

As part of this scheme of a universe cooling down, the four fundamental forces are thought to decouple one by one as random events or through a process of symmetry breaking.

The scheme of the Big Bang given by different cosmologists differs in some details, but the general idea is that energy condensed into mass. During the first phases of the Big Bang photon energy is converted into matter and anti-matter particle pairs and back again into photons. Helium was formed within the first 5 minutes. The universe was opaque plasma until it had cooled sufficiently for atoms of hydrogen to form. When the universe became transparent it entered the phase it is in now where matter is distributed in galaxies and stars, with radiation travelling through space.

Helium
One of the tests of the Big Bang Model is the abundance of helium in the universe. Once helium is formed it is tough to destroy. The Big Bang is thought to have converted 25 % of the matter of the universe into helium. If this is true, no celestial object should have a helium abundance of less than 25 %. Since stars also form helium by thermonuclear reactions the abundance of helium in older stars should always be greater than 25 %.

It has been found that interstellar clouds are composed of 90 % hydrogen atoms, 9 % helium and 1 % heavier elements (see Chapter 11, section 1). According to the Big Bang Model interstellar clouds – the stuff of galaxies – should have 25 % helium, not 9 %, thus there seems to be a problem here.

The universe certainly contains the helium produced by the thermonuclear reactions taking place within stars. Helium was possibly produced at the centres of galaxies when they first formed in the early universe. There may be no helium produced in a Big Bang [14].

Inflation

One of the really big problems with the Big Bang Model is that the universe today does not show the signs of having been produced by an explosion. Explosions are messy. The fallout from explosions is not neat and homogenous. If the universe started in an explosion, one would expect to find temperature differences in different parts of the universe today. In fact, the cosmic background radiation shows only very fine temperature differences apart from which it is equal across the whole universe (we will return to the cosmic background radiation later).

The American physicist, Alan Guth attempted to resolve this problem in 1980 with the Inflationary Big Bang Theory. Guth proposed that the universe evolved from a region much smaller than that proposed by the Standard Big Bang Model (10^{50} times smaller) and then underwent sudden inflation. The minute size of the universe when the explosion occurred allowed light to spread the temperature evenly throughout the universe. Then inflation increased the size of the universe while preserving the uniform temperature. (The size involved is determined by the horizon distance and the speed of light, but I will not describe the technicalities of this here). The theory proposes that a smooth universe was obtained by inflation taking place faster than the speed of light [15].

According to the Inflationary Big Bang Model, at 10^{-43} seconds the universe was less than the size of a proton. The temperature is thought to have been as much as 10^{32} K (one hundred million million million million million degrees). Inflation is said to have occurred between 10^{-43} and 10^{-32} seconds bringing the universe from smaller than a proton to the size of a grapefruit. Then the Standard Big Bang Model resumes. At 10^{-10} seconds after its birth the universe had

---

[14] The helium atom is composed of a nucleus of two protons and two neutrons orbited by two electrons. The neutron is formed by the combination of a proton with an electron with the release of a neutrino under conditions of high pressure.

[15] Guth, Alan H. (1981) A Possible Solution to the Horizon and Flatness Problem. *Physical Review* Vol. 23, pages 348-356.

grown to the size of our solar system. The force which would have caused the universe to suddenly inflate is not named in the theory.

Inflationary Theory was used to resolve the problem of the homogeneity observed in the universe also known as the isotropy of the cosmic background radiation. Inflation produces a smooth universe. In our 'smooth' universe light rays travel parallel in straight lines no matter how far they travel through space; distortion is only detected when light rays skirt around a galaxy – for this reason the universe is described as 'flat' in astronomer speak. Inflationary Theory explained why the geometry of the universe is 'flat'. But then another problem presented itself – how did the rapid inflation stop and the expansion of the universe slow down?

It was calculated that if the total amount of mass in the universe had a critical value, the rapid inflation would slow down due to gravitational forces. If, however, the mass in the universe represented less than the critical value, expansion would be such that galaxies and stars would never have formed and the universe would have flown apart. While if the mass in the universe represented more than the critical value, the universe would have collapsed onto itself in a Big Crunch. It is argued that since the universe has not flown apart and not ended in a Big Crunch, the universe must have the critical value of mass. The problem with this is that this mass cannot be found.

Observation shows that all the stars and interstellar clouds contained in all the galaxies of the universe only represent at most 20 % of the matter that the universe should contain if the theory is correct. Thus, Inflationary Theory requires a new hypothetical form of matter. This invisible and undetectable matter has been named dark matter. (Dark matter will be discussed in Chapter 16). Belief in the existence of dark matter rather than taking the observable universe at face value is a requirement for belief in the Inflationary Big Bang Model.

3. The Multi-Bang

I stuck with belief in some sort of Big Bang throughout most of my time working on the hypotheses in this book. But I knew that the formation of galaxies demanded explanation and a satisfactory explanation had not yet been given. I had various ideas involving the overall structure and dynamics of the universe, but it was only in 2009 that a new idea came to me that there could have been more than one bang. In this case it would be the interaction of motions produced by the various bangs that would create the universe as we know it, including the formation of galaxies.

The basic concepts belonging to the Multi-Bang Hypothesis are as follows:

- The atom is mostly composed of space, and so the creation of atoms explosively creates space.

- God creates subatomic particles each with their own specific properties such as electrical charge. The natural recombination of subatomic particles causes atoms to form.

- Once hydrogen atoms were created at certain points they moved outwards due to the force of explosion. These streams of hydrogen atoms met at the intersections between explosions; their interaction formed eddies, and these eddies became galaxies.

The **Multi-Bang Hypothesis** can be stated thus:

Matter was created in the form of protons tightly packed together at saturation property density. Thus, the matter of the entire universe occupied a small volume. Created electrons were injected into the mass of protons at a myriad of points, possibly over a certain length of time. The high-energy electrons blew apart the proton mass in multiple bangs. Around each point, atoms of hydrogen formed. The natural recombination of a proton with an electron led to the formation of each atom. The space inside each atom became the cause of the cosmic expansion of the universe. The newly-formed atomic hydrogen gas streamed outwards from the points of expansion and met with other streams of hydrogen at intersections between the centres of expansion. The filaments of hydrogen curled round each other at intersections forming eddies; these eddies became galaxies formed from spiral arms of atomic matter.

In the Multi-Bang Hypothesis matter is formed first. Photons or energy come from the transformations of matter. In the Big Bang Model energy comes first and condenses into matter.

One of the inspirations for the Multi-Bang Hypothesis came from watching firework displays. Rockets explode in the sky with lots of points of coloured light moving outwards as expanding spheres. I asked myself, what if there was no gravity and so the points of light were not drawn downwards towards the Earth, but continued to move outwards in increasing circles. What patterns of interaction would they form? This, in my imagination, became an image of the very early universe.

This is the more detailed scenario of the Multi-Bang Hypothesis:

1. The universe was created as a mass of protons packed together. Each proton was composed of two Up quarks and one Down quark. The universe was thus very small, of high density and totally dark. The universe was cold because the protons were not moving.

2. At a myriad of points within this dense mass God created electrons. The injection of high speed electrons moving close to the speed of light broke apart the proton mass.

3. The negatively charged electrons instantaneously recombined with the positively charged protons to form the first atoms – atoms of hydrogen consisting of one proton orbited by one electron. Since most of the atom is empty space, the result was sudden expansion in the form of explosion. (The distance between the nucleus and the first electron shell in an atom is a thousand times the diameter of the nucleus, but in excited states it is much more).

4. Atoms only form at or below 4000 K so the universe had a temperature lower than that of a star such as the Sun at this time.

5. The sudden creation of space inside the atom gave rise to motion which propelled the atoms through external space. Streams of hydrogen atoms started to move outwards from the myriad points of explosion. As atomic matter moved outwards, the four dimensions of spacetime opened up. An expanding universe had been created.

6. The multiple explosions were probably not momentary. Jets of high speed electrons continued over time –possibly millions of years. This gave rise to 'controlled explosion' in which hydrogen was continually supplied from its subatomic ingredients. Controlled explosion gave rise to an isotropic universe.

7. Photons or light energy were emitted from these first atoms. The newly created highly excited atoms may at first have emitted ultraviolet light photons. This light started to move through the universe in all directions. As the universe expanded these photons had their energy diluted and redshifted. The initial ultraviolet light passed through all the visible colours of the rainbow through infrared and finally to the microwave wavelength. It became the cosmic microwave background radiation detected today.

8. The universe was expanding from a myriad of points and hydrogen atoms were flowing outwards. Where the streams of hydrogen atoms met, intersections were formed. These intersections became galaxies composed of atomic hydrogen.

9. The streams of neutral hydrogen became the spiral arms of galaxies. Galaxies behave like eddies –as the spiral arms are drawn in, the galaxy revolves with the axis of spin determined by the dominant streams. Thus, the motion of galaxies originates in the propulsion produced by the initial explosions external to the galaxies themselves.

10. The drawing in of arms of hydrogen gas raises the density of gas at the centre of galaxies. Thus, galaxies often have a central area where the density of gas has raised the temperature giving rise to hot, bright, active galactic cores.

15.1 CHRONOLOGICAL STAGES OF THE MULTI-BANG

1. Dark, cold dense proton mass

2. Electrons created at multiple points

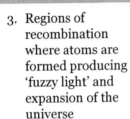

3. Regions of recombination where atoms are formed producing 'fuzzy light' and expansion of the universe

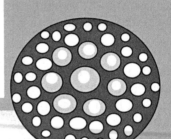

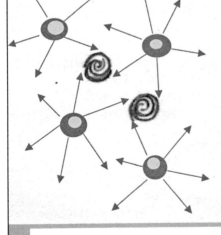

4. Filaments of hydrogen stream outwards and galaxies form at the intersections between filaments and points of expansion

5. Galaxies form with spiral arms of attenuated atomic hydrogen in the outer parts, coalescing into denser arms of atomic and molecular hydrogen towards their centres

What would the size and density of the universe have been at the beginning?

It would be the size of all the protons and neutrons that exist in the universe today i.e. the amount of baryonic matter in the universe today packed together side by side in one locality.

The density of nucleons within atomic nuclei is roughly constant. The constant interior density of nuclei is called the **saturation property**. The density of protons or neutrons in the interior of a nucleus is about 3×10^{17} kilograms per cubic metre, and each proton or neutron takes up about 6 cubic femtometres of volume (Encyclopedia Britannica: Atom).

Protons are composed of quarks. The strong nuclear force sets the distance between the three quarks of which the proton is composed. The strong nuclear force also sets a limit on how far protons can approach each other inside the nucleus. The fixed limit on how far protons can be packed together in a nucleus is the saturation property. (The strong nuclear force was discussed in Chapter 14, section 4). Observations of atomic nuclei show that this limit is never broken.

I take the density of the nucleus – the saturation property – as an indication of the density of the initial universe before space had opened up. From this data the size of the initial universe could be calculated, but not by me. The initial universe would be small and of high density. It would not, however, be crushed down to a single point as the Big Bang Model proposes – this would not be allowed by the saturation property.

How would the protons stay together?
According to the Multi-Bang Hypothesis the state of the universe at the beginning was as a solid localized collection of protons. Protons are positively charged so they do not hold together in the nucleus unless neutrons are present to neutralize the electrical charge. However, I believe that the collection of protons at the beginning held together because they were cold –having a temperature of absolute zero at which all molecular motion stops (see footnote[16]). Therefore,

---

[16] At absolute zero a substance would be so cold that its atoms would cease moving. Absolute zero itself has never been reached either in nature or in the laboratory, although scientists have come close. The coldest atoms of dark

these initial protons were motionless and for that reason they did not repel each other. As soon as energy was added to the proton mass, the protons became heated and were set in motion, and in these conditions they fly apart.

The strong force that holds together the nucleus of atoms is very strong at low energies. This would mean that if the initial universe was composed only of protons, but these protons were cold and had no motion (no vibration and no rotation), the absence of motion would allow them to hold together. Conversely, the strong force is weaker at higher energies. This means that the proton mass could be blown apart by high energy electrons. From this point onwards the protons would be in motion.

How many bangs took place?
Each explosion marked a point of expansion in the universe. At the intersections between explosions galaxies formed. A type of network pattern would be formed.

The number of galaxies in the universe would be related to the number of explosive points. There would probably be more galaxies than explosive points. However, there may have been a reduction in the number of galaxies existing today compared to the initial number due to galactic cannibalism giving rise to fewer, but larger galaxies. Today the universe contains 500 000 million galaxies.

I have proposed that God created the basic subatomic particles. The creation of electrons may have been on-going for millions of years to give rise to 'controlled explosion' (An example of a controlled explosion is the combustion engine. See footnote[17]). Once positively and negatively charged subatomic particles had been created, the formation of atoms would occur by recombination as a chain reaction. A chain reaction is a natural process, occurring over time and

---

clouds in space have temperatures a few degrees above absolute zero. It might be impossible to get to absolute zero, and even if we did we might not know it because no thermometer could measure it (Baker 2007, page 40).
[17] In the internal-combustion engine a fuel-air mixture is reacted converting it into hot gases. The high pressure of the gaseous medium produced by confining it within the engine is used directly to drive the moving parts of the engine. The internal-combustion engine works by the expansion of gases driving mechanical motion.

determined by the conditions of temperature and pressure in the universe.

This means that when the injection of high speed electrons stopped or their supply was used up, the formation of new hydrogen would come to an end, and the formation of new galaxies would cease. Thus, galaxy formation would have occurred over a certain period of time during the early universe and then stopped.

4. Quasars

Curious objects have been found far away at the distances of distant galaxies that resemble bright stars. One type emits a radio signal and was called a quasi-stellar radio source. For each of these there are 20 objects with a similar optical appearance but no radio emission called quasi-stellar objects (QSOs). They are now collectively known as quasars.

Quasars have high redshifts of emission lines from hydrogen atoms. This means that the ultraviolet light they emit when redshifted is seen as blue light. (See footnote[18] on redshifts). High redshift shows that the quasar is far away and that it existed in the universe at an earlier time. A quasar with a redshift of 3 is seen when the universe was one-eighth of its present age and one quarter of its present size. Quasars have even been found with redshifts of about 4.5.

This means that we look back across 90 % of the history of the universe to a time about a thousand million years after the beginning (Gribbon & Rees 1991, pages 47-48). The most distant quasar detected by X-rays is at 15 000 million light-years. A light-year is the distance light can travel in one year. A quasar 15 000 million light-years away is seen as it was 15 000 million years ago. It is the tremendous brilliance of quasars that allows them to be observed at distances of more than 10 000 million light-years.

---

[18] Astronomers measure distances to galaxies and quasars in terms of redshift. This is the displacement of lines in the light spectra from hydrogen of these objects compared with the wavelengths at which these lines occur in the laboratory here on Earth. Redshift results from movement or from the expansion of the universe.

Quasars are brighter than the most luminous galaxies and yet they present the point-like image of a star. Galaxies, on the other hand, appear as fuzzy patches of light because they consist of many individual stars. A quasar can be up to a thousand times more luminous than a giant galaxy containing a thousand billion i.e. a million million stars. Even so, calculations involving light intensity have shown that quasars are less than 1 to 2 light-years across. X-ray emission has shown that the central region is the size of the solar system, and yet the energy release occurs at a rate greater than 10^{12} Suns (a million million times the energy released by the Sun).

Quasars have jets of electrons travelling nearly at the speed of light. This is called synchrotron radiation. Currents of electrons emit electromagnetic radiation, and when the electrons move at relativistic speeds i.e. those approaching the speed of light, the brightness of the radiation increases enormously. Many of these jets are observed to be one-sided. Astronomers claim that the jets are two-sided, but only one half can be seen.

Motion detected in the jets of five quasars, if taken at face value, lead to the conclusion that the electrons are moving faster than the speed of light. Motions that appear faster than light are called superluminal motions. The quasar 3C 273 has a jet of electrons that appear to be moving at a velocity of 10 times the speed of light. Astronomers think that this must be an optical illusion since if it were so it would break one of the laws of the Theory of Special Relativity which states that nothing can move faster than the speed of light (see footnote[19] on

---

[19] Light always travels at the same speed of 300 million metres per second in a vacuum. Light waves do not add their velocity together. The speed of light appears the same for any observer no matter how fast they are moving. This led Albert Einstein to devise his Theory of Special Relativity in 1905.

If the speed of light is constant, then something else must change to compensate. Space and time must distort to accommodate the different viewpoints of observers travelling close to the speed of light. Nothing exceeds the speed of light. For this to be so, distances must shrink and time slow down to compensate. Objects are prevented from travelling at the speed of light because according to $E = mc^2$ their mass increases reaching infinity at the speed of light. The closer an object gets to the speed of light, the heavier it becomes making any further acceleration impossible. Photons do not have mass so they are unaffected.

Special Relativity). The superluminal effect is thought to be caused by the almost head-on orientation of the jet to our line of sight. The electrons are thought in reality to be moving at 99.5 % of the speed of light.

The electron jets from quasars stretch across 10 million light-years of space in some cases. This means that the quasar was active for 10 million years if the electrons were moving close to the speed of light. If the electrons were moving 10 times faster than the speed of light, their source would have been an instantaneous injection.

More than 2000 quasars have been identified. Quasars were more numerous and more luminous about 10 thousand million years ago. Most are found beyond 5 thousand million years ago. The brightest quasar, 3C 273, lies at a distance of two thousand million light-years from the Earth. The distribution of galaxies throughout the universe is due to their on-going existence. Quasars, on the other hand, are distributed in the far reaches of the universe which means that they existed early on. The lack of near-by quasars is an indication that they have had limited lifetimes, and have now ceased to exist.

Supermassive black holes?

The currently accepted theory is that quasars are the nuclei of active galaxies viewed from different angles. The synchrotron radiation is thought to be produced by supermassive black holes at the centres of these galaxies. A black hole capable of producing the amount of energy observed for quasars would have to have a mass of 100 million solar masses. It would be swallowing stellar material from an accretion disk into a horizon or Schwarzschild radius measuring 2 AU. The material would radiate energy as it spiralled into the black hole due to acceleration. Twin jets of synchrotron radiation would emerge out of the black hole from the poles of the axis of spin. It is claimed that often only one of the jets is seen because the nearest one to the line of sight is seen, while the one on the other side is not seen or is very faint. It is also claimed that the stars of the active galaxy that is the host to the quasar are not seen as one would expect as a disk surrounding the nucleus, because the galaxy is too far away.

Thus, an object approaching the speed of light would contract in length, age more slowly and become heavier.

I accept that there are active galaxies with bipolar jets of high speed electrons emanating from the core. An example of an active galaxy with bipolar jets is Cygnus A. The source of the energy of this galaxy is probably collision with another galaxy. There must have been many galactic mergers over time with the production of synchrotron radiation. The explosion of stars within galaxies as supernovas is also a source of synchrotron radiation. However, I think that these examples of synchrotron radiation involving galaxies are separate categories of phenomena from quasars.

Black holes have been proposed as the source of quasars because only black holes are small enough to fit the bill while having the theoretical capability of producing the massive amounts of energy observed for quasars. However, I do not think that black holes are the correct explanation for quasars.

Power-engines of the universe

I propose that quasars are the power-engines of the universe. Matter in the universe would have been created at the locations of quasars from electron jets and protons recombining to form atomic hydrogen. This would explain the tremendous energy associated with quasars in the early universe.

If electrons cannot move at or faster than the speed of light, the Multi-Bang may not have been instantaneous. Electrons may have been injected into the universe over a period of time producing a controlled explosion. Quasars have synchrotron radiation consisting of electrons that appear to be moving close to the speed of light and quasars show an existence through time, although they no longer exist now.

Jets of high speed electrons associated with quasars are often observed as single jets; this is what one would expect if electrons had been injected into the universe.

The things observed about quasars makes them likely candidates for the Multi-Bang beginning.

By certain techniques it has been found that a fuzzy haze sometimes surrounds a quasar nucleus. The Multi-Bang Hypothesis would interpret this as representing the region where atoms were formed by recombination producing a fuzzy type of light within the cloud.

166

Some observations concerning quasars show connections between quasars and galaxies:

- Some quasars have nuclear jets of ionized gas as well as electron jets.

- 30 % of nearer quasars (with a redshift of less than 0.6) appear to be interacting with a nearby galaxy. Quasars sometimes appear to be connected to galaxies by filaments of atomic hydrogen.

- Quasars have often been found to lie amongst clusters of galaxies. 3C 206 is surrounded by 20 faint galaxies.

The above observations could be explained by the Multi-Bang Hypothesis as streams of atomic hydrogen emanating from the zone where atoms could have been formed around quasars, with filaments of atomic hydrogen moving outwards to form galaxies at the intersections between the points of expansion.

Quasars found at the heart of clusters of galaxies is what one would expect to find if the Multi-Bang Hypothesis was correct.

A link could be explored between the total energy of expansion of the universe (excluding the notion of Inflation) and the energy emitted by quasars.

According to the First Law of Thermodynamics the total amount of energy in the universe is conserved.

It is proposed that at the beginning of the universe, matter was created as protons packed together in one locality with a temperature of absolute zero. There was no motion or energy in the universe at this moment. High speed electron jets at quasar points injected energy into the universe and expansion occurred due to the space in the atom.

If a quasar is a thousand times more luminous than a giant galaxy and there are 500 thousand million galaxies in the universe (although not all galaxies are giant galaxies), there may have been 500 million quasars.

167

One could calculate the amount of energy released by 500 million quasars to see if it was in the same order as the total kinetic energy of expansion of the universe. This would show whether quasars are good candidates to fit into the scheme.

5. Cosmic background radiation and cosmic rays

The cosmic background radiation (CBR) in the universe is composed of photons. It is a fossil of the initial moments of the universe. Intergalactic space is traversed by photons of light; it is also traversed by cosmic rays. Intergalactic cosmic rays are composed mainly of accelerated protons. I believe that cosmic rays constitute another background to the universe. Both of these backgrounds lend support to the Multi-Bang Hypothesis.

Cosmic background radiation

The universe today is uniformly permeated by microwave radiation with a temperature of 2.735 K or −270.4 °C. Thus, it is just above absolute zero which is 0 Kelvin. This is the cosmic background radiation (CBR). It is detected by radio astronomers in intergalactic as well as galactic space. Although it is cold now and in the form of soft microwaves, these photons are thought to be a fossil of the very early universe when it was hot and filled with strong ultraviolet light. The expansion of the universe has caused these photons to be cosmologically redshifted from a very short wavelength to a long wavelength.

The cosmic background radiation is always cited as one of the proofs of the Big Bang, especially since it was predicted that the light from the Big Bang would be found in the universe today. In fact, the cosmic background radiation does not come from the primordial fireball described by the Big Bang Model when the universe is conceived as originating as pure energy at millions of degrees temperature in a blinding flash of light. The cosmic background radiation comes from an era in the universe when matter had formed into atoms and the temperature of the universe was 3000-4000 K. This is the temperature of the surface layers of a cool star. This era is thought, according to the Big Bang Model, to have existed some 300 000,

500 000 or million years after the initial flash. According to the Big Bang Model it would be at this time that the cooling of the universe would allow matter and radiation to be 'decoupled' and matter to release radiation as a blackbody[20].

The homogeneity of the cosmic microwave radiation indicates that the matter from which it was emitted was very evenly distributed. There were no galaxies at this time, or even hints of matter clumping together. The cosmic background radiation has a redshift of $z = 1000$. This means that the universe at this time was smaller by a factor 1000 in linear dimensions which means that its average density was a thousand million times greater than it is now (Gribbon & Rees 1990, page 58). That the cosmic background radiation has the same temperature in all directions shows that there was an extremely smooth distribution of matter in the universe at this time.

It has been calculated that for galaxies to have formed as they are observed today, there should have been greater densities by 1 part in 200 at the time $z = 1000$. The isotropy of the cosmic background radiation shows that there could not, in fact, have been irregularities greater than 1 part in 20 000 at $z = 1000$ (Gribbon & Rees 1990, pages 58-59). This is one of the major problems for the Big Bang Model in explaining the formation of galaxies.

Background radiation now redshifted into the microwave frequency does appear to be a remnant from the early universe. However, I do not think that it proves the Big Bang Model for the origin of the universe; on the contrary, it seems to cause some intractable problems for the Model.

For the Multi-Bang Hypothesis the formation of atoms can be taken as time zero, the beginning of the universe itself. Atoms would have been formed throughout the universe and upon their creation released ultraviolet light whose remnant is the cosmic background radiation detected today. At the same time, explosive expansion sent matter moving in all directions throughout the universe as the universe unfurled.

---

[20] A blackbody radiator is a solid, liquid or gas that absorbs radiation at all wavelengths and emits radiation at all wavelengths. The spectrum of the blackbody depends only on its temperature. The cosmic background radiation came from a blackbody. The blackbody in this case was neutral hydrogen atoms.

Recent observations show the cosmic background radiation to have hotter spots. The variation is only about a hundredth of a percent difference in the background temperature (Zeilik 2002, page 469). According to the Multi-Bang Hypothesis, the points of creation of atomic matter would be expected to be hotter than the intervening regions in the cold proton mass. This might explain the pattern of slight hot spots within the cosmic background radiation found today.

Therefore, according to the Multi-Bang Hypothesis, the homogeneity or isotropy of the cosmic background radiation could be explained by the formation of atoms at many centres throughout the universe. These atoms would release photons that would be seen to be coming from all parts of the universe in all directions.

The myriad points might produce homogeneity while the discreet identity of the points might have produced the slight patchiness of hotter and cooler spots in the cosmic background radiation.

The expansion of the universe from a myriad of points, rather than from one point would be the explanation of the cosmological principle –that the universe does not have a centre and space is the same in all directions. Explosion from one point as the Big Bang Model proposes would not produce a universe looking like this one. (It has, however, often been stated that this intuition is wrong).

Cosmic rays

Cosmic rays are subatomic particles travelling at high speeds. They come from two sources, one source is the Sun and the other is outer space.

Solar cosmic rays are protons and electrons blasted from the Sun's surface by solar flares. There is a large outflow of material from the Sun forming a plasma of particles which travel out into the solar system far beyond Neptune (at 30 AU) to as far as 100 AU[21]. The plasma of particles is known as the **solar wind**; it also constitutes the Sun's corona; and they are known as solar cosmic rays. These particles may have a temperature up to 500 000 K.

---

[21] AU stands for astronomical unit which is the distance between the centre of the Sun and the Earth. It is about 150 million kilometres or 93 million miles.

The Earth and other planets deflect the solar wind by magnetic fields that trap particles and form a magnetosphere around the planet. The deflection of solar cosmic rays by the magnetic field is fortunate for us. Cosmic rays break chemical bonds and ionize the atoms they collide with by direct impact –this could produce lethal mutations. A large number of cosmic rays hitting the Earth would be detrimental to life but we are protected by the Earth's magnetic field.

Solar wind passing by the Earth contains on average five particles per cubic centimetre, but the numbers vary depending on solar flare activity. It mainly consists of protons which are the nuclei of hydrogen atoms. The velocity of these particles is on average 500 kilometres per second. These are low-energy cosmic rays.

There are also cosmic rays from outer space travelling at a much greater velocity and some of these reach the Earth's atmosphere. The cosmic rays observed at sea level are not the primary particles, but rather secondary radiations produced by interactions between the primaries and the top of the atmosphere. Some of the secondary radiations occur as showers of photons or electrons. The primary cosmic rays arriving at the top of the Earth's atmosphere consist predominantly of positively charged particles.

Primary cosmic rays consist mostly of protons, but there are also heavier nuclei. Cosmic rays are 84 % protons; 14 % helium nuclei stripped of both electrons called alpha particles; and 2 % electrons. The small number of heavier nuclei are in the carbon-nitrogen-oxygen range, with some iron nuclei and a few more exotic particles such as heavy quarks. The relative abundances of elements in the primary cosmic ray population reflect the abundances of these elements in the universe.

The < 2 % cosmic rays that are heavier nuclei including carbon, nitrogen, oxygen and iron are ionic species produced by the collision of cosmic ray protons with these elements in interstellar molecular clouds. These ionized elements come from interstellar dust grains. The heavier nuclei having lost their electrons are accelerated by the impact and the total velocity of the products of the impact equal the velocity of the original proton impactor.

Accelerated carbon-12 nuclei occasionally collide with a fast-moving proton or alpha particle in the interstellar medium. This can cause the

171

carbon nucleus to lose a proton by nuclear fission and become boron-11. The production of the light elements lithium, beryllium and boron by cosmic ray impacts is called spallation. These light elements also form part of the cosmic ray population.

The cosmic rays from outer space are extremely energetic. They are not travelling at 500 km/second like the solar wind, but often at over 87 % the speed of light (Encyclopedia Britannica: cosmic ray). The current thinking is that cosmic rays are produced by high energy events within the galaxy such as supernova explosions; the particles are then accelerated by spiralling in magnetic fields that wind themselves through the spiral arms of the galaxy.

Let us examine the speeds involved:
The material ejected from Type I supernovas typically has a velocity of 10 000 km/second and the material ejected from Type II supernovas has a lower velocity of 5000 km/second (Zeilik 2002, page 366). Supernova 1987A which exploded recently in the Large Magellanic Cloud was a Type II supernova with an expansion rate measured by the Doppler redshift of 17 000 km/second (Zeilik 2002, page 368). Novas are lesser, but more frequent explosions in the galaxy. The Firework Nebula nova of 1901 produced material expanding away at 1200 km/second (Zeilik 2002, page 363).

These speeds are comparable with the speed of rotation of gas clouds at the core of spiral galaxies where black holes are thought to lurk. Active galaxies such as Seyfert galaxies attain speeds of 10 000 km/second in their highly active cores.

It is often stated in astronomy texts that cosmic rays travel close to the speed of light. The speed of light is approximately 300 000 km/second. A cosmic ray travelling at 87 % the speed of light has a velocity of 261 000 km/second. There is a very large difference between this velocity and 10 000 km/second. Thus, it is difficult to see how a particle expelled from a supernova at a velocity of 10 000 km /second could be accelerated to a velocity of 261 000 km/second or more.

There are further enigmas with cosmic rays –they do not come from any particular direction. Supernova events can be located, so if supernovas were producing the cosmic rays it would be expected that their supernova source would be revealed by the direction the cosmic rays were travelling in. However, cosmic rays have an isotropic distribution. It is argued that there must be some kind of scattering mechanism within the galaxy that diffuses the rays. Stray magnetic fields associated with moving clouds have been suggested as a scattering mechanism. Cosmic rays are thought to have been travelling through the galaxy for millions of years before reaching the Earth.

I suggest that highly energetic cosmic rays had their origin at the original creation of matter. I have proposed that the original creation of matter occurred around quasars. Quasars have jets of electrons moving close to the speed of light, thus they have the energy range capable of the acceleration observed in cosmic rays.

Cosmic rays are protons accelerated to 87 % the speed of light or 261 000 km/second. These protons may have come from the original proton mass (described as part of the Multi-Bang Hypothesis) and been accelerated by the explosive events that launched the universe into existence.

These primary cosmic rays would, if this were true, have been travelling through intergalactic space for thousands of millions of years before by chance entering a galaxy.

Quasars found outside galaxies and from an earlier era in the history of the universe are a credible acceleration mechanism given the energy ranges. Recent supernovas of dying stars within galaxies do not have the necessary power to explain the phenomenon.

The isotropic distribution of cosmic rays suggests that they form a background, just as the cosmic background radiation of photons forms a background. A background of primary cosmic rays would also be a fossil from the origin of the universe.

The composition, speed and distribution of cosmic rays are strong support in favour of the Multi-Bang Hypothesis since the hypothesis proposes a form of matter and energy levels that match what is observed in cosmic rays.

The remnants of the early universe are found *outside* galaxies. In this category I would place the cosmic background radiation of photons, quasars with their synchrotron radiation of high speed electrons and cosmic rays of accelerated protons. More recent events of the universe are found *inside* galaxies –these include the formation of molecular clouds, on-going star formation and nova and supernova events.

Cosmic rays do not give off light except upon collision, when a gamma ray may be released. Cosmic rays travelling through intergalactic space are unlikely ever to collide with anything since there is nothing much to collide with. Cosmic rays in intergalactic space, if they exist, would be difficult to detect since they do not give off light.

Although primary cosmic rays travelling in intergalactic space cannot be seen, it is possible that they are felt – as the darkness of outer space in the night sky (Re: Olbers' Paradox).

Thus, primary cosmic rays would be the original protons caught up in the expansion of the universe and accelerated in all directions by the explosive events of the Multi-Bang. Maybe high temperature prevented them from forming atoms, and so they continued to travel through the universe in all directions giving rise to an isotropic distribution. At a later stage, when they entered galaxies some cosmic rays collided with helium, carbon, nitrogen, oxygen and iron elements in interstellar clouds, and these accelerated nuclei joined the cosmic ray population.

Therefore, the extremely high velocity of these positively charged cosmic ray particles cannot be explained as originating in supernovas –much more powerful events are indicated. I propose the cosmic ray background as evidence for the explosive events of the Multi-Bang at the beginning of the universe.

7. Black holes –an end to spacetime

The current version of the Big Bang Model conceives the universe as having originated in a singularity – an object the size of a single proton. Current thinking conceives black holes as singularities. The Big Bang is seen as the beginning to spacetime, while black holes constitute an end to spacetime.

If black holes were simply dense objects rather than singularities, and if they continued to obey the laws of physics, then they would bring about an end to spacetime in what would resemble the beginning proposed by the Multi-Bang. Let us take a look into the unseen.

In accordance with the Theory of General Relativity, where large masses of matter exist in the universe, space is pinched in. Stars and galaxies produce curvature in spacetime. Taken to the extreme, gravity is not just the curvature of spacetime, but the elimination of spacetime in a black hole.

A star larger than three solar masses, when thermonuclear reactions cease, ends its life as a black hole (this was mentioned in Chapter 13, section 3). Supernova explosion slams matter in the core of the star together and the core implodes. When a certain density of matter is reached gravitational collapse becomes inevitable. It is expressed in terms of the gravitational radius.

The gravitational radius is also called the Schwarzschild radius after the German astronomer Karl Schwarzschild who in 1916 predicted the existence of collapsed stellar corpses that emit no radiation. Another name for this is the event horizon. The size of the gravitational/ Schwarzschild radius/event horizon is thought to be proportional to the mass of the collapsing star. For a black hole of 10 solar masses, the radius would be 30 km (*Encyclopedia Britannica 2011 Standard Edition*: Black hole).

The collapsing core of a massive star becomes so dense, with such a strong gravitational field, that even light does not escape from it. Photons go into orbit around the black hole object because the escape speed from the star is greater than the speed of light. Since no light is emitted from it, the black hole is black.

The original name for black holes was frozen stars. Any object passing too close to a black hole will spiral in and this includes light. Once it passes the event horizon, matter or light plummets in and is not seen again since the event horizon marks the radius where the escape velocity exceeds the speed of light. According to General Relativity time stops at the event horizon to the outside observer, and hence the star is frozen in time. Time ends where light does not travel.

The strength of the gravitational field in black holes causes gravity to crush the atom such that atoms become degenerate. Degeneracy in its ultimate expression is when the atom collapses and the space inside the atom is removed with only subatomic particles remaining. A supernova explosion that heralds the formation of a black hole releases electrons as synchrotron radiation and energy in the form of neutrinos. Thus, the collapse of the atom is accompanied by the expelling of energy. What remains at the end is a stellar corpse formed of dense solid neutrons.

The view that black holes constitute singularities where spacetime ends involves the idea that the black hole object continues to collapse down until it becomes a mathematical point. This mathematical point known as a singularity would have zero volume and therefore infinite mass.

According to the Singularity View, the gravitational radius of a black hole is the radius below which the gravitational attraction between particles of matter causes them to undergo irreversible gravitational collapse. This would mean that inside the gravitational radius there is "no solid stuff", only infinite mass.

The alternative view of black holes is that the core of the massive star despite its original massive size collapses down until it is only a few kilometres across. It then ceases to collapse. This would mean that the Schwarzschild radius or gravitational radius represents the size of the black hole object and it would contain extremely dense, solid stuff. The Schwarzschild radius is directly dependent on the mass of any object. In the hypothetical situation that the Sun became a black hole, it would have a radius of 3 km, and if the Earth became a black hole, it would measure 1 cm radius (figures provided by Zeilik 2002, page 379).

After much reflection, I finally reached the view that the Singularity Model of Black Holes is wrong since I do not believe that matter can collapse down until it has no volume; I think that a barrier to further collapse must be hit. Since formulating this opinion I have discovered that the barrier to further collapse is called the saturation property. The density of atomic nuclei consisting of protons and neutrons is constant; nuclei do not become denser than the saturation property of 3×10^{17} kilograms per cubic metre.

If black holes consist of the protons and/ or neutrons of collapsed atoms tightly packed together, they would be subject to the strong nuclear force and have the same saturation property as the nuclei of atoms. This would mean that a black hole is a highly dense, small object, but not a singularity. It would also mean that black holes continue to obey the laws of physics, rather than constituting a breakdown in the laws of physics.

Over time galaxies may become littered with stellar corpses including black holes. These dark lumps of matter are the end points to spacetime. In black holes the atom is crushed and space is removed, leaving only the subatomic particles. Where gravity has squeezed out even the space inside the atom, space ends, and where space ends, time also ends.

If the original state of the universe was of tightly-packed protons, then the dark lumps of matter known as black holes where spacetime ends have much resemblance to the initial state. Gravity brings about the clumping together of matter – if the Multi-Bang Hypothesis is a true representation of the beginning of the universe, then gravity returns matter to something resembling its original state.

Conclusion

Up to the end of the 19th century and beyond, it was assumed that the universe had started out as dispersed gaseous matter that had clumped together to form stars and the planets of our solar system. The universe was generally assumed to be infinite both in extent and in duration. It was not known that there was anything beyond our own galaxy except maybe infinite space, since the lights seen in the night sky which are galaxies, were taken to be stars in our own galaxy.

We have come a long way since then. Scientific investigation has revealed a universe that is vast, but finite in extent and old but finite in duration. The universe had a beginning where all the energy of the universe was concentrated into one place.

That the universe had an explosive beginning and ever since has been expanding and cooling down is consistent with the Hot Big Bang Model and with the Multi-Bang Hypothesis.

In the Big Bang Model the universe starts off as pure energy in a single point at thousands of millions degrees. Light becomes matter and antimatter, and later on mainly matter is left.

In the Multi-Bang Hypothesis asymmetrical matter is created, not photons. There is no antimatter, and light is produced from the high energy states of matter. The initial universe is small, but not tiny or 'infinitely small', and hot in having an overall temperature of about 4000 K. It is at this initial time that the cosmic background radiation was released.

Quasars appear to be the beasty that a Multi-Bang enthusiast would hunt for. Quasars play a key role in the Multi-Bang Hypothesis since they represent star-like points with massive amounts of synchrotron radiation composed of electrons moving almost at the speed of light. The injection of energy into the initial universe would cause a controlled explosion.

When the initial proton mass (the total mass of the universe) was blown apart by high-speed electrons, this led to the formation of the first atoms of hydrogen. The places where the first atoms formed became multiple points of expansion in the universe. The space created inside the atom was the cause of the expansion of the universe.

Atoms of hydrogen set in motion streamed out in filaments in all directions, and where streams in motion met, galaxies formed. Thus, energy and motion created space in between matter. Moving matter creates spacetime –and this is the universe.

The Multi-Bang Hypothesis explains the formation of galaxies by the newly-formed filaments of atomic hydrogen, following the directions of expansion meeting at intersections and curling around each other to become giant eddies of matter. Galaxies are the subject of Chapter 16.

The detection of microwave cosmic background radiation is given as one of the proofs of the Big Bang Model, even though the Model has a hard time explaining the isotropy of this background radiation. The Multi-Bang Hypothesis has less problems explaining why this radiation comes from all directions and is very homogenous since light would have been released from a myriad of bangs.

The Multi-Bang Hypothesis has observational proof in the form of cosmic rays travelling close to the speed of light. These accelerated protons would have been set in motion by quasars at the initial creation of the universe and continue in their trajectories in all directions. The Multi-Bang Hypothesis predicts that a population of cosmic rays moving in all directions will one day be detected in intergalactic space.

The Multi-Bang Hypothesis provides a link between the very small – the formation of atoms and the very large –the unfurling of the universe with its mega-structures. The expansion of the universe is determined by the space inside the atom.

The 'empty space' of the atom is not empty; it contains motion. Subatomic particles have intrinsic motion in terms of spin – intrinsic motions give subatomic particles their properties such as the force of magnetism. The atom is created by the orbital motions of electrons around the nucleus.

The universe was created by the expansion outwards of matter composed of atoms. The interaction of large numbers of atoms formed galaxies. Stars formed in molecular clouds orbiting the centres of galaxies. Stars spin on their axes and the planets orbiting stars spin also. The three dimensions of space and the dimension of time were created by motion on a cosmic scale.

Atoms interact by exchanging electrons and forming molecules. Atoms and molecules communicate by emitting and absorbing photons of light.

Light and motion creates a universe. Gravity brings matter back together. Eventually gravity squeezes out space and brings an end to motion. Planets are engulfed within stars when stars die, and galaxies merge with each other. In stellar corpses named black holes the space inside the atom is crushed out. Motion inside the atom stops, space is gone and time stops. Gravity returns matter to its initial amassed state and darkness.

The Multi-Bang theory is based on motion and the interaction of matter in motion creating the universe. God set up the universe and determined the total amount of matter that it would contain. God created protons. God injected energy into the universe by creating

179

electrons moving close to the speed of light. These initial creative acts launched into existence a universe that could undergo cosmic evolution as it continued its existence through time.

Bibliography

Baker, Joanne (2007) *50 Physics Ideas You Really Need to Know* Quercus

Cassé, Michel (2003) *Stellar Alchemy: The Celestial Origin of Atoms* Cambridge University Press

Encyclopedia Britannica 2011 Standard Edition: Black hole; Cosmic ray; Cosmos; Gamow, George; Gravitational radius; Hawking, Stephen W; Hubble, Edwin Powell; Internal-combustion engine; Lemaitre, Georges; Olbers'paradox; Quasar; Steady-state theory.

Gamow, George (1947/1961) *One, two, three, infinity.* Viking Press

Gribbin, John & Martin Rees (1991) *Cosmic Coincidences: Dark Matter, Mankind and Anthropic Cosmology* Black Swan

Hawking, Stephen W. (1988) *A Brief History of Time: From the Big Bang to Black Holes* Bantam Books

Maxwell, James Clerk (1876) *Matter and Motion* Society for Promoting Christian Knowledge. Reprinted in 2002 by Prometheus Books.

Sagan, Carl (1980) *Cosmos* Random House

Shaw, Andrew (2006) *Astrochemistry: From Astronomy to Astrobiology* John Wiley & Sons Ltd.

Zeilik, Michael (2002) *Astronomy: The Evolving Universe* 9th edition Cambridge University Press

GALAXIES: EDDIES IN THE UNIVERSE

Introduction

The description given of the structure and types of galaxies in this chapter shows that although stars in our galaxy and the other galaxies both appear as points of light in the night sky to the naked eye, they are, in fact, very different categories of phenomena demanding quite different explanations for their genesis and continuing evolution.

This chapter follows on from the last chapter to show how the Multi-Bang Hypothesis explains the formation of galaxies.

At the end of Chapter 16, the currently very important subject of dark matter and dark energy will be examined since dark matter is often invoked to explain the dynamics of galaxies.

1. Galaxies

Ninety nine percent of matter, mostly consisting of hydrogen and helium, is found within galaxies. Cold, low-density interstellar hydrogen in galaxies is formed of atoms. There are also molecular clouds composed of molecules of hydrogen where the density of hydrogen is higher. Molecular clouds also contain clouds of silicate dust associated with carbon molecules and water. About half the hydrogen in our galaxy is atomic, the other half molecular. Most of the mass of our galaxy is in stars; a much lower mass is in atomic and molecular hydrogen gas clouds. The stars of galaxies are plasmas of dissociated atoms of hydrogen with the consistency of water. Main sequence stars are surrounded by regions of hot ionized hydrogen. Most of the matter in galaxies is contained within spiral arms.

The size of galaxies varies from a few thousand to a few hundred thousand light-years across. Galaxies can occur alone or in large groups. The distances separating galaxies are of the order of one million light-years. All galaxies are rotating.

Galaxies are moving away from each other. The universe is expanding by 5-10 % every thousand million years (Hawking 1988, page 49). There was a time when galaxies must have been close together.

Galaxies are observed as patches of light. The light of a galaxy is emitted by stars and clouds of hot gas. Galaxies have on average a hundred thousand million stars. There are well over 500 thousand million galaxies in the universe.

The Milky Way

Our galaxy, the Milky Way is about 98 000 light-years across. So, a rocket, even if it were travelling at the speed of light would take 98 000 years to get from one side to the other. The engines of the Space Shuttle rocket produce a velocity of 3630 metres per second, compared to the speed of light which is 300 000 kilometres per second, so a rocket with the present level of technology would take 8000 million years to cross our galaxy. This gives an idea of the size of the galaxy.

The Milky Way is composed of five spiral arms that revolve about the centre every 250 million years in the plane of the galaxy. It has 400 thousand million stars mostly located in the spiral arms. Our solar system resides within the Orion arm about 30 000 light-years from the centre. The average density of the Milky Way is one hydrogen atom per cubic centimetre. The Milky Way is a spiral galaxy also known as a disk galaxy. It belongs to a cluster of about 30 galaxies called the Local Group.

16.1 THE MILKY WAY – Top View

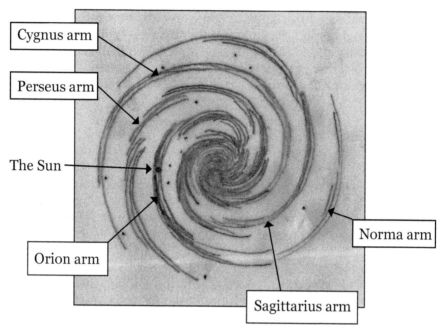

(This illustration was redrawn from a schematic model of our galaxy in Zeilik 2002, page 390).

2. How did galaxies form?

This is one of the great questions of science. Whereas stars in galaxies continue to form, the galaxies themselves are thought to have been formed at a remote epoch and new ones are not thought to be forming now.

The problem for the Big Bang Model is how to get galaxy formation in an expanding universe with evenly distributed matter. Galaxies could only have formed when the universe cooled down enough for matter and radiation to be separate (4000 K). Expansion has the effect of diluting matter. If gravity is the controlling force, how could gravity clump together this diffuse matter to form discreet galaxies while expansion continued?

There are many theories that purport to explain how the density of matter became slightly higher in some areas or how the expansion slightly lagged behind the general expansion in certain areas. The result of the proposed mechanisms −none of which has been very convincing −is the formation of clouds of hydrogen and helium. These enormous clouds are thought to have collapsed under the pull of their own gravity to form embryonic galaxies.

Therefore, according to the Big Bang Model, between one and two thousand million years after the initial explosion, clouds of gas equivalent to the mass of a hundred thousand million stars underwent gravitational collapse contracting inward due to gravitational instability. In the process of contraction these clouds fragmented into smaller clouds. The smaller clouds underwent further gravitational collapse to form the stars of which galaxies are composed.

The rotation of galaxies is explained by the Big Bang Model by the initial clouds having a tiny speed of rotation on their own axis. When these clouds underwent gravitational collapse and contracted in size, they started to spin faster and faster due to the conservation of angular momentum. The disk shape of galaxies occurred because as a galaxy spins on its axis of rotation, the initial spherical shape of the galaxy underwent centrifugal forces that flattened it into a flat disk shape.

The above account summarizes the current thinking on the problem of galaxy formation. The explanations given are not very satisfactory. The Multi-Bang Hypothesis describes a new way of thinking about the formation of galaxies.

As explained in Chapter 15, the Multi-Bang Hypothesis proposes that matter was formed in localized areas identified as possibly corresponding to quasars, and that these areas became the centres of expansion increasing the size of the, at one time, very small universe. Galaxies formed through the interactions of matter in motion at intersections between the points of expansion. Galaxies became the places where matter accumulated.

Galaxies are essentially eddies in the universe – places where streams of atoms like cords become curled around each other. These streams of atomic hydrogen became the spiral arms of galaxies.

The rotation of galaxies occurs because galaxies are eddies in the universe formed by matter in interacting motion. The axis of rotation would have been determined by the motion of the dominant filament or filaments of matter of which a galaxy is composed gaining over that of the smaller filaments.

The role of gravity

The universe began with explosion –in that we are all agreed. In the current theory, gravity is the cause of almost everything in the universe, whereas in the Multi-Bang view it is not.

According to the current theory, galaxies were formed by gravitational collapse in giant clouds of hydrogen. According to the Multi-Bang Hypothesis galaxies were formed by matter in motion, and the motion came from multiple explosions external to the galaxies themselves.

According to the Multi-Bang model of galaxy formation, matter accumulates at the centre of galaxies raising the density of the core of galaxies as the spiral arms of matter are drawn in. Thus, in the Multi-Bang model, gravity plays a part in raising the density of matter within galaxies, but more importantly the motion of galaxies originates from outside the galaxies themselves.

With the Multi-Bang Hypothesis the formation of galaxies is explained in terms of motion rather than gravitational collapse. Motion is fuelled by energy.

Special Relativity

Mass has an associated energy that can be released if the matter is destroyed. Matter releases energy by nuclear fission and nuclear fusion. The equivalence of mass and energy was established by Einstein in a mathematical footnote to the Special Theory of Relativity in the famous formula: $E = mc^2$

$E = mc^2$ Where energy E is equal to a quantity of matter with mass m times the speed of light c squared. In this equation, c acts like an exchange rate between matter and energy.

When matter is converted into energy it produces motion. The three dimensions of space and the dimension of time were created by matter in motion. In the scheme I am proposing, the mega-structures of the universe were also produced by motion. In this scheme, gravity, although crucial, has a more limited role than in the current scheme and motion has a greater role.

3. Star populations

Stars are of different ages since they are continually being formed within galaxies, and they have different compositions in terms of the proportion of heavy elements in their surface layers. This section describes the official classification of populations of stars.

Population I and Population II stars

Stars are formed in the spiral arms of galaxies where atomic hydrogen has been converted into clouds of molecular hydrogen. This subject was discussed in Chapter 13, section 1. Stars undergo an evolution through various stages. They start as protostars enshrouded in dust; they mature into main sequence stars whose thermonuclear reactions cause them to shine; and they end their lives as stellar corpses, the larger ones exploding as supernovas that send shock waves through the galaxy.

Nearly half the stars of the universe are binary stars, which orbit around a common centre of mass. Some are very close and some far apart. Triple stars are also common. The simplest explanation for this is the capture of one star by another. Double stars are often similar in mass. If this were not the case and the stars were very different in mass, the smaller star would soon be drawn into the larger star and disappear.

Stars are labelled as Population I stars or Population II stars depending on their age. Stars that are now old because they formed when the galaxy was young are labelled Population II stars. Stars that are young i.e. they have existed for less time are labelled Population I stars. Population I and Population II stars are distinguished by metallicity. Metallicity means the proportion of heavy elements in a star compared with the light elements of hydrogen and helium. Heavy elements include carbon, oxygen, silicon and iron etc. They are not all metals, but the term metallicity is used because the presence of iron is sampled as an indication of the overall heavy element content.

It has been found that Population II stars that were formed early on have very little heavy elements (0.01-0.02 %), while Population I stars that were formed more recently have 2-3 % of their mass in heavy elements. Thus, the age and the heavy element content of a star are inversely proportional. The Sun has 2 % heavy elements.

Stars are thought to produce heavy elements within their cores by nucleosynthesis when they reach old age and the red giant phase. When stars are sampled for metallicity, it is the surface layers of the star, not the core that is sampled. According to current theory stars acquire their metallicity by being formed in clouds that themselves contain a large proportion of heavy elements; the heavy elements came to be located in the dust clouds by being blown out of stellar cores by supernova explosions. The higher metallicity of younger Population I stars would be produced by the on-going production of heavy elements enriching the molecular clouds of galaxies over time. Metallicity is not a measure of the heavy element production of a star itself.

In Chapter 12 I suggested that these heavy elements may not have been produced by the cores of dying stars. I proposed that the fundamental elements such as carbon, oxygen, silicon and iron were created as catalysts or living cells associated with dust in dark molecular clouds. If this were the case, the metallicity of stars would be due to these heavy elements being drawn into the surface layers of the star. At certain stages in their life, stars do appear to draw in large quantities of dust from surrounding dust clouds. Stars also engulf planets that approach too close. These in-puts would raise the metallicity of the surface layers of the star. It is not clear how the drawing-in of heavy elements by stars would be related to age or if it would be more influenced by location. We will see shortly that metal-poor stars are located in the halo, while metal-rich stars are located in the spiral arms of the disk and in the centre of the galaxy.

Open clusters and globular clusters

Stars are found in groups called star clusters. Clusters are aggregations of stars that move through space together as a unit. There are open clusters and globular clusters of stars. Open clusters have less than a hundred to a maximum of a thousand stars at low density. The star density is not more than a few stars per cubic light-year. The Milky Way has some 20 000 open clusters. In open clusters the lower-mass stars follow the main sequence on the H-R diagram with some higher-mass stars having become red giants.

Globular clusters have a spherical shape that appears like a fuzzy ball to a small telescope. They have many stars densely packed together. A globular cluster has as many as a hundred stars per cubic light-year and a total of up to a million stars of one solar mass or less. The stars of globular clusters do not follow the main sequence on the H-R diagram for the most part and many stars have evolved off the main sequence into the red giant stage. This indicates that globular clusters are very old, even 13 000 million years old. The Milky Way has about 200 globular clusters.

Open clusters have Population I stars that are often bright and bluish-white in colour because they are hot, young stars. Globular clusters have Population II stars that tend to be reddish in colour because they are cooler, old stars. The Sun and many moderately hot stars like it, and faint red dwarf stars are, however, classified as Population I stars. Population I stars have a range of ages between a few tens of millions of years to ten thousand million years old. It is thought that the stars of each cluster formed at the same time, but vary due to having different masses.

Galaxy structure

Studies of our own galaxy show it to be composed of three main parts: a central region called the nuclear bulge, a disk with spiral arms and a halo.

- The nuclear bulge of the galaxy contains the nucleus, the very heart of the galaxy. The nuclear bulge is shaped like a flattened ball containing old, metal-rich Population I stars.

- The halo is spherical and encircles both the nuclear bulge and the disk. It contains globular clusters of very old, metal-poor

Population II stars. The globular clusters orbit the galaxy at all angles, some with highly elliptical orbits.

- The disk contains open clusters of young metal-rich Population I stars. It also contains clouds of interstellar gas and dust that extend out further than the furthest stars. The disk has several spiral arms that extend for thousands of light-years and contain most of the matter of the galaxy. In the spiral arms of our galaxy about half the matter is in neutral hydrogen, molecular hydrogen and dust clouds, and the other half has formed stars.

Stars are divided into Population I and Population II stars on the basis of high or low metallicity, however, in reality there is a continuous range of abundances between 3 % and 0.1 % heavy elements. Thus, there are not simply two clear-cut groups. There is, however, a clear correlation between metal abundance and average distance from the galaxy's disk. The lower the metal abundance, the further the stars are found from the plane of the disk (Zeilik 2002, page 405-406). The metal-poor stars inhabit the halo.

16.2 THE MILKY WAY – Side View

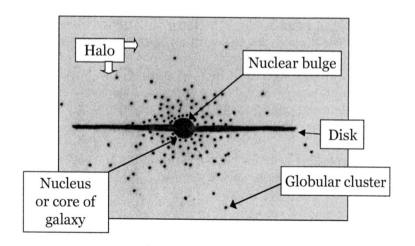

(This illustration was redrawn from a schematic model of our galaxy in Zeilik 2002, page 390).

The oldest stars in the galaxy are found in the halo. The halo is thought to be the fossil remains of the birth of the galaxy. Globular clusters of the halo follow highly elliptical paths. The disk of the Milky Way is 98 000 light-years in diameter, but the globular clusters of the halo fly out as far as 300 000 light-years from the galaxy's nucleus. This is an indication that our galaxy had an original size of 300 000 light-years diameter and over time has been drawn in.

The oldest stars in globular clusters in the halo are some 14 000 million years old (Zeilik 2002, page 422), while the youngest objects in the galaxy are found in the disk.

4. Galaxy type and age

The universe is populated by galaxies of different shapes. There are elliptical, spiral, and irregular shaped galaxies.

- The elliptical galaxies show a variety of shapes, some spherical, some elongated and some oblate like a squashed sphere. They tend to have a roundish shape. There are dwarf elliptical galaxies as well as giant and supergiant ellipticals.

- The spiral galaxies show the form of flattened disks. There are also disk shaped galaxies with no spiral arms called lenticular galaxies. Of the brightest galaxies in the sky, over 60 % have been found to have well-defined spiral arms. Surveys show that both nearby and distant galaxies have spiral structure (Zeilik 2002, page 398-399). This shows that spiral structure endures over time.

- The irregular galaxies have no spiral structure or symmetry. Dwarf irregulars are very common.

Galaxies are classed I, II, III, IV, and V, depending on their luminosity. Galaxies of class I are the most luminous and galaxies of class V the least luminous. It has been found that the most luminous galaxies are larger in size. Class I galaxies are supergiants. The luminosity class of a spiral galaxy is also correlated to the form of its spiral structure. In class I, highly luminous galaxies the arms are long and well developed and have a high surface brightness; in class III galaxies the arms are patchy and fuzzy; and in class V galaxies there may be barely a hint of spiral structure (*Encyclopedia Britannica 2011 Standard Edition*: Cosmos: Classification of galaxies).

All galaxies rotate, but those spinning most rapidly are the spiral galaxies. The elliptical galaxies do not rotate much at all and thus have a very small angular momentum. Spiral galaxies have a higher angular momentum in the disk than in the halo. The more spin a galaxy has, the flatter its disk shape. Irregular galaxies have no shape to speak of and therefore the motions within them do not have an overall direction. This gives an overall angular momentum of zero.

The irregular galaxies contain large quantities of gas, the spirals less and the ellipticals have almost no gas at all. Star formation occurs at different rates in different types of galaxy. Star formation in galaxies depends on the amount of gas available.

Irregular galaxies have a rotating disk of neutral hydrogen gas. They have about 15 % of their total mass as atomic hydrogen. Dust clouds are not conspicuous. Hot young stars with a bluish colour are forming from this gas and creating H II regions of ionized hydrogen.

Spiral galaxies have about 5 % neutral hydrogen. They have young stars in the disk and old stars in the halo and nucleus. The nuclei of spirals appear reddish because they contain older stars. Therefore, the progression from the bluer irregular galaxies to redder spiral galaxies reflects the galaxy's stellar populations.

Both dwarf and giant elliptical galaxies have less than 1 % neutral hydrogen. In these galaxies nearly all the gas has been used up and star formation has ceased. Ellipticals are populated by old stars. Ellipticals tend to appear redder than spiral galaxies. Calculations involving the mass and luminosity of galaxies indicate that elliptical galaxies have a greater proportion of low-mass stars with low luminosities (main sequence stars of class M). They are also likely to contain a high proportion of objects that contribute to mass but not to luminosity such as stellar corpses in the form of neutron stars, black dwarfs and black holes. They may also have many small failed stars such as brown dwarfs.

Varying age of galaxies

The dominant opinion at the moment is that all galaxies have the same age, and they all formed 10 000 million years ago. However, the speed of rotation of galaxies, the amount of neutral hydrogen gas they contain and the age of stars within the galaxies suggests that galaxies do, in fact, have different ages; they were not all formed at the same time.

The elliptical galaxies could be classified as old and the irregular galaxies as young with the spiral galaxies in between. The following is a possible scenario:

Elliptical galaxies would be the first galaxies to form in the early universe. The universe at this time would have been small in extent and replete with newly produced atomic hydrogen moving at high speed. The galaxies that formed at intersections between the points of expansion would contain clouds of gas and later stars moving at high velocity. Objects moving at high velocity in a gravity field have elliptical rather than circular orbits (this will be discussed in Chapter 17 of Part III). Somehow this elliptical signature remained impressed on the shape of elliptical galaxies. Although elliptical galaxies show little rotation, the constituent stars often have high individual velocities.

Some elliptical galaxies are enormous in size and classified as supergiants. They may have achieved this size because at the time that they were formed, large quantities of matter were close at hand for their formation since the early universe was much smaller. Some may have increased their size by galactic mergers (this will be discussed shortly).

Elliptical galaxies are now old and are inhabited by old stars of reddish colour. They have used up their supplies of neutral hydrogen gas and have no new star formation.

Spiral galaxies would have been first formed in a slightly later era when the universe had grown in size. The matter moving through this universe, although abundant, would not have had the energy that the matter first produced had. Filaments of this matter collected together as galaxies with spiral arms.

The first stage in the formation of a galaxy would involve matter collecting together at an intersection from all directions in three dimensional space. The galaxy would form from uneven contributions of atomic hydrogen gas from different sources entering the galaxy from different directions. At this stage the galaxy would not have a dominant direction of motion. This would give the galaxy an irregular morphology. At this early stage the galaxy may have a very active core.

Over time the dominant stream of matter would win over the other streams and an axis of rotation would be established for the galaxy. As the galaxy continued to rotate, a pattern would be formed causing the galaxy to become a flattened disk with most matter held close to the plane. When a galaxy becomes disk shaped, its spiral arms become apparent. The spiral arms may have been there before, but in 3D rather than in a clearly visible 2D shape.

The stars born during this evolution preserve, in their trajectories and dynamics, the characteristics of the galaxy at the time of their formation. For this reason the older stars –those of the globular clusters in the halo –show a quasi-spherical arrangement and trajectory around the galactic centre; while the brighter and younger stars of population I stars are mostly confined to the galactic disk.

Therefore, elliptical galaxies were the first type of galaxy to form in the early universe, and they are now the oldest galaxies. At a later era other galaxies formed with an irregular shape when first formed and a disk shape with visible spiral arms after a certain lapse of time. Irregular galaxies are either young galaxies or newly-produced galaxies from the galactic merger of older galaxies.

Zeilik (2002, page 426) states that spiral galaxies appear redder than irregular galaxies. He also states that within the spiral group, a galaxy appears redder as the nuclear bulge grows larger and the spiral arms become less extensive.

The observation that the central mass of a galaxy is seen to grow larger as the spiral arms grow smaller is evidence that the spiral arms are being drawn in. It has also been observed that there is a correlation between the tightness of the spiral windings and the size of the central nuclear mass. This dynamic observed in galaxies strongly supports the Multi-Bang Hypothesis for galaxy formation.

As stated earlier, in very luminous galaxies the spiral arms are well defined, while in duller galaxies they become indistinct. The luminosity class, as well as being related to size, could also be related to age. The spiral arms disappear in older galaxies.

Although the different types of galaxy may be of different ages, galaxies are no longer being formed now. There would have been an age of galaxy formation probably divided into two phases; an early phase producing elliptical galaxies some of them giant in size, when the universe was smaller and more energetic with abundant matter available, and a later phase producing irregular galaxies which evolved into spiral galaxies when the universe had grown in size and had more dispersed matter.

The creation of matter in finite quantity at the beginning of the universe and the expansion of the universe in size would mean that the formation of new galaxies would cease completely at a certain point in time.

5. Interactions between galaxies

Galaxies are not randomly dotted about the cosmos; they are associated with each other in clusters that are gravitationally bound. They also appear to form superclusters that give a structure to intergalactic space.

In their movement through space, galaxies may encounter other galaxies. Large galaxies may capture smaller galaxies and consume their contents. This is called galactic cannibalism.

Clusters of galaxies

Galaxies are found in clusters that orbit a common centre of gravity. The Milky Way belongs to a cluster of over 20 galaxies called the Local Group. The Local Group cluster is dominated by the Milky Way and Andromeda galaxies with the centre of gravity at a mid-point between these two.

The Large Magellanic Cloud and the Small Magellanic Cloud are galaxies that belong to the Local Group. They are both connected to our galaxy by a bridge of hydrogen gas. A large, thin envelope of neutral hydrogen and a thread of stars also connect the two Clouds. The Large Magellanic Cloud and the Small Magellanic Cloud are both irregular galaxies with no nucleus and smaller than the Milky Way. It is not yet fully understood how galaxies came to be joined by filaments of matter.

The Local Group cluster is moving towards the massive Virgo cluster.

A typical cluster contains about a hundred galaxies and is separated some tens of millions of light-years from its neighbouring clusters. The spacing of galaxies in a cluster amounts to only about a hundred times a typical galaxy's diameter. This means that, relatively speaking, galaxies in a cluster are very crowded together. The result is that galaxies pass close by each other.

Galactic cannibalism

The massive galaxies – supergiant ellipticals are at least 10 million times bigger than the smallest galaxies – the dwarf ellipticals. So, it is easy to understand how the massive galaxies disrupt the smaller ones by gravitational attraction, and how the larger ones devour the smaller ones in what is known as galactic cannibalism.

There is evidence of galactic cannibalism in that the supergiant elliptical galaxies often have extensive halos; multiple nuclei near their centres; and are located at the centre of clusters. These elliptical galaxies may have increased their size through galactic cannibalism. Supergiant elliptical galaxies tend to define the gravitational centres of clusters of galaxies.

Close encounters between galaxies pull out tails of matter from the galaxies, and bridges between galaxies are thought to come from the same interaction. Galactic mergers can induce strong bursts of star formation that convert interstellar clouds into stars at up to a hundred times the rate seen in other galaxies. Starbursts are seen in irregular galaxies.

Cygnus A is an example of a double-lobed source of synchrotron radiation. The bifurcation of this giant elliptical galaxy is thought to have occurred when it recently swallowed a spiral galaxy.

Galactic mergers may constitute a proof of the Multi-Bang Hypothesis in that the fact that galaxies collide shows that they are moving from points of intersection in all directions; they are not moving outwards from a single point. The Big Bang would have occurred at one point, whereas the Multi-Bang would have involved many points.

Superclusters

Clusters of galaxies appear to be grouped into superclusters. On the grand scale, clusters of galaxies appear to form a web-like pattern made of chains and knots of galaxies. The overall pattern appears to be of superclusters with large holes and voids in between.

When galaxy formation ceased, the fundamental pattern of superclusters would have been fixed. However, the fundamental pattern could be modified by galactic cannibalism reducing the number of galaxies over time, while giving rise to giant galaxies. The orbiting of galaxies around giant galaxies or around common centres of gravity within clusters could also modify the overall pattern.

Cooling flows

Streams of hot gas have been found flowing through intergalactic space towards the centre of clusters of galaxies. These flows of matter have been found because they emit X-rays. They are known as cooling flows because the hot gas is radiating energy and therefore losing energy and cooling down (Gribbin & Rees 1991, pages 144-145).

Cooling flows contain very large amounts of matter that may contribute matter to galaxies at the heart of clusters and they are also associated with individual galaxies that are not at the centre of clusters. This flowing of matter in intergalactic space is not fully understood.

Comment

It is possible that the streams of hot gas flowing towards the centre of clusters of galaxies known as cooling flows are remnants of the filaments of atomic hydrogen that, according to the Multi-Bang Hypothesis, became galaxies. The existence of threads of hydrogen gas in intergalactic space strongly supports the Multi-Bang Hypothesis.

As far as the web-like pattern of superclusters is concerned, the Multi-Bang Hypothesis would predict that the large voids correspond to centres of expansion (or former centres of expansion), while the peripheral chains and knots of galaxies correspond to regions of accumulation of matter at intersections between centres of expansion.

6. Galactic centres

There is the idea today that all galaxies have massive black holes at their centres. Supermassive black holes would have to range in size from a few million to a few hundred million solar masses to account for the energy involved in the high velocity motions of matter at the centres of galaxies.

Sagittarius A

The centre of our galaxy is a very hot, active place. The core cannot be seen by optical telescopes because it is obscured by thick clouds of dust. But radio telescopes that detect radio and infrared emissions show that large quantities of matter are moving at speeds of several hundred kilometres per second around the centre. Old Population I stars are tightly packed into the nuclear region and there is a large H II region of ionized hydrogen.

The region that marks the centre of the Milky Way is called Sagittarius A. At the centre, inside Sagittarius A, lies a radio source less than 13 AU (2 light-hours) in diameter that gives off synchrotron radiation of high speed electrons spiralling in a magnetic field. It also produces gamma rays and X-rays. It has been proposed that the synchrotron radiation comes from the supernova explosion of stars in the core.

Sagittarius A has a spiral structure with ionized hydrogen gas of a few million solar masses rotating about the centre at high speed. Beyond this there is a molecular cloud with hydrogen molecules at 2000 K and other molecules such as carbon monoxide. The molecular cloud forms a ring of gas and dust outside the ionized region that goes out as far as 30 light-years and forms a disk shape in the plane of the galaxy.

Radio and infrared observations show rapid rotational motions near the Milky Way's core. The rotational velocities increase closer to the core. These rotational movements demand an explanation.

Massive black holes

One explanation is that there is a black hole at the centre of the galaxy providing the mass and gravitational attraction that drives the movement of clouds around the centre. The speed of rotation would require a mass of 4 million solar masses in a region only 0.1 light-years across (Zeilik 2002, page 403). Other galaxies have also been observed to have a similar mass concentration in their nuclear region. The idea that a supermassive black hole lurks at the heart of our galaxy and other galaxies is, however, yet to be confirmed.

If there is a black hole at the centre of our galaxy, its mass indicates that it must have been formed by more than one star. Other galaxies also appear to have black holes at their centre and the size of the black hole is related to the size of the galaxy. The black hole mass found at the centre of galaxies appears to be a thousand times smaller than its surrounding galaxy. The central object can be hundreds of times more massive than the mass at the centre of our galaxy.

Eddies

In accordance with the Multi-Bang Hypothesis for the formation of galaxies, one would expect there to be very high density at the centre of the 'eddy'. The spiral arms with the matter that they carry would be drawn into the centre of the galaxy with the impulse of their motion coming from outside the galaxy itself. This would result in high velocity at the centre combined with a high concentration of matter in the form of gas clouds and stars at the centre.

The young stars of the galactic arms would be drawn into the centre over thousands of millions of years and be old by the time they reached the centre. Thus, the central nuclear region has old Population I stars, and many of these may have undergone supernova explosions and become black holes. That the mass of the black holes coalescing at the centre is proportional to the total mass of the galaxy supports the Multi-Bang Hypothesis for the formation of galaxies.

Active galaxies

Our galaxy is classified as a quiet galaxy, but there are others classified as active galaxies. The gas at the centre of active galaxies is in a disturbed state. One type of active galaxy is a Seyfert galaxy. The cores of many spiral galaxies resemble Seyfert galaxies in that they are very active. Seyfert galaxies are generally spiral galaxies with a bright nucleus surrounded by a bluish ring produced by star formation. The gas in the nucleus of Seyferts moves at the very high speed of around 10 000 kilometres per second. The source of this energy is not known.

One theory is that the nucleus of an active galaxy is occupied by a single supermassive black hole. To produce the amount of energy observed, the black hole would have to accrete matter at a rate of two solar masses a year (*Encyclopedia Britannica 2011 Standard Edition*: Cosmos: Black hole model for active galactic nuclei). A competing theory holds that the energy source results from multiple supernova explosions in a confined space near the centre of the active galaxy.

Comment

There is a relationship between the mass of a galaxy and the mass at its centre; the mass of a galaxy is a thousand times greater than the mass of its central object.

The central object may, in fact, be a conglomeration of black holes left by the supernova explosions of old spent stars found in the central nuclear region of the galaxy.

The existence of a proportion between the mass of the central region of a galaxy and the overall mass of a galaxy lends support to the Multi-Bang Hypothesis of galaxy formation; this mode of galaxy formation involves the drawing-in of filaments of atomic hydrogen as spiral arms towards the centre of a galaxy. On the other hand, there is no reason why a supermassive black hole at the centre of a galaxy would show a relationship to the overall mass of the galaxy in which it operated.

7. Dark galaxies

Apparently there are galaxies which do not shine because they contain no stars or hot gases. They are found very far away and existed in the early universe. They were discovered through observations of quasars which existed in an even earlier universe.

Lyman-alpha series
Large clouds of atomic hydrogen gas have been found in intergalactic space unassociated with any visible galaxies. They have been detected because they lie as foreground objects to distant quasars.

One way of detecting these clouds is by the Lyman-alpha series of lines for the spectrum of hydrogen. The Lyman-alpha series is produced by light from a quasar passing through and being absorbed by clouds of cold atomic hydrogen gas that lie between us and the quasar.

Galaxies are generally only studied in detail out to a redshift of 0.3. This covers only a few thousand million years of cosmic history and a few thousand million light-years of space. The Lyman-alpha lines provide information about the regions of the universe beyond a redshift of 1.7. The dark clouds of hydrogen highlighted by quasar light often have redshifts exceeding 3. This light comes from the universe when it was only one to four thousand million years old.

Gravitational lensing
Another way of detecting large clouds of cold atomic hydrogen is by their gravitational lensing effect. The light from some distant quasars appears as multiple images. It has been puzzling that in some cases of multiple images, there is no trace of a bright galaxy in the right place to produce the lensing effect. For multiple images to be produced, the light must pass through more than one lens. Consequently, one can surmise that some large dark objects are doing the lensing.

The detecting methods show that a typical dark cloud is as big as a small galaxy 35 000 light-years across. The mass of a dark cloud may be between 10 million and 100 million solar masses. Some dark clouds may be 10 000 million light-years away from us.

The matter referred to above is baryonic matter which means that it is composed of atoms containing nucleons made from quarks.

Most of the matter in the universe is hydrogen which has one proton and one electron. However, in estimating the amount of baryonic matter in the universe the mass of electrons is ignored because the mass of the electron is 1800 times less than that of the proton. Thus, the mass given for all the matter in the universe is the mass of all the protons and neutrons in the universe.

Critical value for the total mass

Views about the universe changed when the Standard Big Bang Model was modified by the Inflationary Big Bang Model proposed by Alan Guth in 1980. If one assumes that the universe underwent a phase of sudden inflation for a few split seconds at the beginning, then one needs a mechanism to halt the run-away expansion and slow it down again. The proposed mechanism is gravity acting upon the total mass of the universe and putting the brakes on. For this to work in the equations, a critical value for the total mass of the universe is required. There is a problem –the universe only contains 10-20 % of this critical mass in baryonic matter.

The argument goes that with less than the critical value of mass in the universe, the universe would continue to expand so rapidly that matter would simply fly apart and would never have clumped together to form galaxies; galaxies exist, therefore, the critical value of mass in the universe must exist. Where is the undetected matter? The answer given is that 80-90 % of the universe consists of invisible dark matter.

Open, closed and flat spacetime

In accordance with the idea that the universe has a critical value for its total mass, the universe is often described as 'flat'. Flat is the finely balanced point between an open universe and a closed universe. If the curvature of spacetime is open, the universe expands forever, while if the curvature of spacetime is closed, the universe recollapses upon itself. The balance point between these two alternatives is flat spacetime where the universe expands but slows down in its expansion. Since the universe is not undergoing rapid inflation now, it is said that we must live in a universe where spacetime is flat.

The word 'flat' is also used by astronomers to de⸱ light rays traverse the universe in parallel lines as⸱ travelling along a flat plain. Only occasionally do⸱ curved path around a galaxy. If the universe was f⸱ matter, light rays would be scrambled by deflection⸱ impossible to see out into the universe. The fact tha⸱ generally deflected is taken to indicate that the unive⸱ appears to be an observation pointing away from the⸱ matter).

Rotational curve of galaxies

The idea of the existence of dark matter comes from beli⸱ Inflationary Big Bang Model, but it has been reinforced b⸱ observations of the motions of galaxies and clusters of gal⸱ rotational curve of galaxies is deemed to show that galaxie⸱ contain more matter than can be seen.

In the solar system the outer planets take longer to complet⸱ orbits and move more slowly than the inner planets. That th⸱ speed of planets decreases with the square root of their distai⸱ the Sun is one of the Newtonian laws of gravity.

The rotational curve of a galaxy is the graph of its rotation spee⸱ different distances from the centre. The surprising result of wel⸱ studied spiral galaxies is that as one goes out from the centre, th⸱ rotational velocity rises to a constant value and then maintains tł⸱ value for as far out as measurements can be made. In a spiral dis⸱ galaxy, the stars farther out take longer to orbit the centre of the⸱ galaxy, but they do not move more slowly. Thus, stars sweep arour⸱ the central nucleus of galaxies regardless of their distance from the⸱ centre and the velocity of the outer stars is not less than that of the⸱ inner ones.

If galaxies have been formed by gravitational collapse, then they are only obeying the laws of gravity if they contain invisible halos of dark matter providing additional mass. The observed rotational velocities of gas clouds in the outer parts of spiral galaxies are so large that it is thought that they would not be bound to the galaxies at all unless the galaxies were more massive than the observations of galaxies indicate. It has been proposed that galaxies contain five times more dark matter

bright stars and this provides the gravitational force that boosts
rotation of the outer parts of a bright galaxy.

...xies orbit around each other in groups called clusters. The
...ive speeds of different galaxies in a cluster indicate how much
...er the cluster contains. The amount of matter needed to make
...ters gravitationally bound is often calculated to be ten times more
...n that of the observed visible matter. Hence the presence of dark
...tter is invoked to explain the formation of clusters of galaxies.

...what could dark matter consist of?

...on-baryonic dark matter

...ost of the theories about the composition of dark matter involve
...on-baryonic stuff. Non-baryonic dark matter would not be made up
...f atoms or atomic nuclei since it does not absorb or emit light.

One candidate for dark matter is neutrinos with a small mass of 20 eV
(Gribbon & Rees 1990, page 130). Neutrinos flood the galaxy after
supernova explosions and they carry large quantities of energy, but
like photons they have not been found to have any mass. It is
hypothesized that if neutrinos were produced by the Big Bang and if
they had a small mass they could constitute the missing dark matter.

Other candidates for dark matter are the hypothetical particles of
Supersymmetry such as neutralinos, axions and photinos. These
particles have to be massive, that is to say, at least the size of protons
or they would not contribute enough mass to make up the 80-90 %
missing mass in the universe. They also have to be very weakly
interacting or they would already have been found. Experiments have
been set up underground to catch WIMPs (weakly interacting massive
particles), but none have been detected so far.

The reason for favouring exotic particles as candidates for dark matter
is that more tangible forms of matter never represent a large enough
amount to constitute all of the dark matter that theory requires.
However, there are forms of 'normal matter' that are dark in that they
do not emit light.

There are no traces of lines corresponding to anything except
hydrogen. There are difficulties detecting helium so the helium
content is unknown, but there are no heavy elements associated with
these dark clouds or dark galaxies.

If clouds like this exist in isolation in space, one would expect that they
would disperse fairly rapidly. The question has been asked, what
holds them together?

I propose that these galaxy-sized clouds of cold atomic hydrogen are
proto-galaxies formed by matter in motion as described in the Multi-
Bang Hypothesis. They would represent the first stage in galaxy
formation. I propose that dark galaxies are dark because star
formation has not yet taken place in them to switch on the
thermonuclear lights. Dark galaxies are located very far away, because
they represent galaxies which existed in an early stage of the universe.

8. Dark matter and dark energy

Dark matter and dark energy are key ingredients of current
cosmological theories. Their influence upon the universe is deemed to
be opposite – the first is the cause of most of the gravitational effects
in the universe, while the second counters and negates gravity.

Dark matter and dark energy are both hypothetical forms of matter
that have never been detected.

Dark matter

Until the 1980s cosmologists believed that the universe is composed of
the matter that can be seen. The matter in the universe that can be
seen forms stars and galaxies that shine as lights. Stars and galaxies
shine because the matter of which they are composed either emits
light or reflects light; even 'black bodies' shine[22].

---

[22] As mentioned before, a 'black body' is called by this odd name because it
only radiates and does not reflect light. An example of a black body is the
Sun.

Baryonic dark matter

First of all, there is matter that has failed to ignite as stars. This includes tiny stars called brown dwarfs that are too small for thermonuclear reactions to have got started. They are warm but not very bright and so are difficult to detect except by their gravitational effects on the larger star they orbit (Brown dwarfs were described in Chapter 13, section 5).

The other major type of dark 'normal matter' is stellar corpses. Dead stellar remnants such as white dwarfs cool down over time and become black dwarfs. They are called black dwarfs because these dead stars are cold and emit no light. A black hole is the stellar corpse of a more massive star. Black holes count as baryonic matter since they are not 'holes' but highly dense bodies. (This category does not include hypothetical primordial black holes or mini black holes that are said to have been formed by the Big Bang out of non-baryonic matter). There are also neutron stars that must eventually run down their energy.

New stars are born so long as a galaxy has supplies of hydrogen gas from which to form them. Over time the gas runs out and star formation stops. As stars progress towards old age, the galaxy is likely to become littered with stellar corpses that no longer emit light. The galaxy may become faint, emitting less light and redder light, but its mass will remain in the form of baryonic dark matter.

The planets of our solar system can be seen since they reflect the Sun's light and are nearby. The planets of other planetary systems cannot be seen from Earth and therefore constitute a type of dark matter. There may be very large numbers of planets orbiting other stars in our galaxy, but they would not contribute any significant percentage of mass to the galaxy.

Asteroids, meteorites and comets are forms of dark 'normal matter' since they are invisible except when seen very close up. This larger grade material cannot be seen at a distance because it does not emit light or interact with light. Telescopes see dust and molecules; they do not see lumps of material.

To this list of baryonic dark matter I would add dark galaxies which are composed of cold atomic hydrogen clouds the size of small galaxies. Dark galaxies do not have stars and so do not emit light.

They have been detected through their absorption of quasar light. Dark galaxies have been called failed galaxies (Gribbon & Rees 1990, chapter 8), but they could be proto-galaxies that will later on produce stars.

I would also suggest the possibility of intergalactic cosmic rays as constituting part of dark 'normal matter'. Cosmic rays are highly accelerated protons (mostly) that do not emit light unless they collide with other atoms. Collision is a rare event inside the galaxy, thus the cosmic rays that reach Earth are thought to have been travelling for millions of years before being detected here. Collision would be even less likely in intergalactic space, so a cosmic ray population could form a type of unseen dark matter in outer space.

Therefore, although forms of normal baryonic matter such as failed stars and stellar corpses can be termed dark matter –they are invisible to telescopes being too dim to be seen –their calculated upper quantity does not resolve the theoretical problem of the missing mass in the universe; they do not constitute the dark matter that cosmologists are searching for.

Dark energy

The Inflationary Big Bang Model requires dark matter to put a stop to the initial rapid inflation of the universe. This produced the belief that the universe has a critical total mass and is poised in an expansion which is slowing down. Spacetime is 'flat' in that the universe is neither flying apart in run-away expansion, nor ending in a rapid Big Crunch.

Thus, the theory predicted that the expansion of the universe should slow down over time and this cosmic deceleration should be detectable. However, when observations of distant galaxies became possible, this prediction was not borne out.

Hubble's Law

The founder of extragalactic astronomy was American astronomer Edwin Powell Hubble (1889-1953). Hubble studied nebulae containing stars known as Cepheid variables [23]. Hubble provided the first evidence in 1924 that these nebulae and stars were not located within the Milky Way, but in other galaxies several hundred thousand light-years away. This gave rise to a new vision of the cosmos –it was now seen to contain many galaxies, not just one, and it provided the first evidence of expansion.

Hubble measured the flux of light from Cepheid variable stars and compared it to their luminosity. The distance to the galaxy containing the star could be estimated using the inverse square law for light [24]. It was found that the universe is expanding at a constant rate.

Hubble's Law states that the farther a galaxy is away from us, the greater its radial velocity of recession. The relationship between recessional speed and distance is expressed as Hubble's constant (H). The law is formulated thus:

$v = H \times d$

v is the radial velocity in kilometres per second
H is the Hubble constant
d is the distance in millions of light-years

The radial velocity is thus directly proportional to the distance. The constant has to be determined through calibration with other measuring methods.

More recently, the distance of galaxies further away from us has been estimated by the flux of light from Type 1a supernovas which have very

---

[23] A Cepheid or Cepheid variable is a star which varies in brightness due to regular variation in size and temperature. The prototype is the star Delta Cephei.

[24] The luminosity of a star is the total amount of electromagnetic radiation (including visible light) that the star emits. Flux is the amount of light reaching, for example, one square metre on Earth from the star. Flux is dependent on distance due to the inverse square law for light. Light from the star radiates out into space in a sphere – for this reason the intensity of the light decreases in a way that is inversely proportional to the square of the distance.

high luminosity and therefore can be seen at greater distances. Type Ia supernovas have a uniform luminosity at maximum of about 10^{10} solar luminosities (ten thousand million times brighter than the Sun). These very bright explosions allow distances to galaxies to be estimated as far away as 1.5 thousand million light-years away (Zeilik 2002, page 421).

The method of measuring distances using Hubble's Law relies on obtaining a reliable value for Hubble's constant. The value is obtained from galaxies of known distances, but data does not always totally agree. Data from Type 1a supernovas give a value of between 21 and 19 km/second/million light-years for the constant. Other data obtained from observing Cepheids obtained a value of 23 km/second/million light-years. Zeilik opts to use the value of 20 km/second/million light-years (Zeilik page 421).

Hubble's Law, whatever its actual value, implies that the universe began expanding at a finite time in the past. If Hubble's constant is taken to be 20 km/second/million light-years, then the age of the universe is 15 000 million years old. The oldest stars in globular clusters are some 14 000 million years old – this is Zeilik's justification for taking the value of the constant as 20 km/s/Mly.

A repulsive force

When two separate research teams (one American and one Australian) measured the flux of light from several dozen Type 1a supernovas in very distant galaxies, they found that the light was weaker than expected. After a certain number of deliberations, it was concluded that this apparent anomaly could only be explained by the galaxies being further away than expected.

It was proclaimed that the expansion of the universe is accelerating.

Cosmologists could have used this new data to call into question the Inflationary Big Bang Model and the reality of the undetected dark matter that would prevent acceleration. However, some cosmologists went on to proclaim the existence of a hitherto unknown repulsive force. The repulsive force would be some sort of cosmic antigravity. They named it dark energy.

Those that believe in dark energy claim that the empty space between galaxies exerts a force that has caused universal expansion to accelerate over the past 7000 million years. Dark energy is represented as arising from non-atomic matter whose only effect is a repulsive form of gravitation. Michel Cassé (2003) uses the term quintessence for dark energy. He describes it as a "negative pressure component" (Cassé 2003, page 195).

The believers in dark energy back up their theories with concepts from Quantum Mechanics. Quantum Mechanics – in its philosophical manifestation as the Copenhagen Interpretation of Quantum Mechanics – swept away the idea that empty space is a vacuum, now conceiving it to be seething with virtual subatomic particles popping in and out of existence. Virtual particles could be the stuff of dark energy, while hypothetical particles could be the stuff of dark matter!

The critical density of matter in the universe is calculated to be 10^{-29} g/cm^{-3} if the Inflationary Big Bang Model is correct. This then requires another form of matter causing the universe to accelerate its expansion. For those who adhere to this belief, this indicates that the composition of the universe is mostly dark matter and dark energy with very little atomic matter. Michel Cassé (2003) gives the composition of the universe on pages 200, 207 and 210 as follows:

70 % quintessence (dark energy)
10-30 % gravitating matter (mostly dark matter)
2-5 % nuclear matter (normal atomic or baryonic matter)

On page 195 Cassé (2003) describes the universe as composed of 2/3 quintessence exerting a repulsive form of gravitation and 1/3 non-atomic dark matter. Similar figures are given by other authors.

If this model for the composition of the universe was correct, it would mean that the universe which has been observed with scientific instruments and which has been detected through scientific analyses only represents between 2 and 5 % of what is out there. Cassé declares "the cosmic insignificance of the atom" (Cassé 2003, page 195).

Are the galaxies moving away from us at an accelerated rate?

In fact, the universe is not accelerating in its expansion; it is only accelerating relative to the theory that it should be decelerating.

"The extra stretching it causes of space, **compared to deceleration**, puts the supernovas farther away than expected." (Zeilik 2002, page 423 bold type added).

The galaxies containing the observed Type 1a supernovas are at a distance predicted by Hubble's finding that the universe is expanding at a constant rate. The slowing down in expansion predicted by the Inflationary Big Bang Model with its dark matter subsidiary theory has not been supported by the facts. As to dark energy, it is neither here nor there.

Dark matter and dark energy are not necessary

The hypothesis that I am proposing declares the significance of the atom. The Multi-Bang Hypothesis takes the vast universe of baryonic or atomic matter more or less at face value. It does not require any hypothetical forms of dark matter or dark energy.

Firstly, the universe of the Multi-Bang does not undergo a phase of sudden inflation; therefore no extra matter is required for the stopping of inflation.

Secondly, according to the Multi-Bang Hypothesis, galaxies are not formed by gravitational collapse in giant clouds of hydrogen; galaxies are formed by the intersection of matter in motion curling into eddies. Motion was produced outside the galaxies themselves. In this model one would expect the velocity of clouds of hydrogen and stars in the spiral arms to be just as high in the outer areas as in the inner areas. This is because the impulse of the motion did not come from gravity, but from explosion. Thus, in the Multi-Bang Hypothesis the explanation for the dynamics of galaxies is found in extra-galactic motion, not in extra mass.

With the Multi-Bang Hypothesis the velocity observed in the spiral arms of galaxies is what the hypothesis would predict in a universe formed only of atomic/baryonic matter.

The Standard Big Bang Model has taken on board the idea that most of the universe is composed of non-baryonic dark matter in the form of exotic particles. I strongly dispute this view of the universe. I do, however, believe in the existence of baryonic dark matter that may take the form of black holes, black dwarfs, brown dwarfs, planets, dark galaxies and extragalactic cosmic rays.

210

As far as expansion is concerned, in the Multi-Bang universe, expansion comes from multiple centres, not just one. This may affect the form of expansion in terms of acceleration.

It would appear that the total mass of the universe is not high enough to impede, by the action of gravity, the continued expansion. This means that we live in a universe with open spacetime. The continuing expansion of the universe is what is observed, as would be predicted by the Multi-Bang Hypothesis. No form of dark energy is required to explain this expansion rate.

Conclusion

In this last chapter of Part II, I have tried to summarize what is known about galaxies; their structure, the star populations they contain, their patterns and behaviour.

The features of different types of galaxies suggest that they were not all formed at one time, and are of different ages, although galaxies are no longer being formed now.

I have proposed a mechanism by which galaxies were formed that is radically different from the present understanding. The explanation of galaxy formation given by the Multi-Bang Hypothesis is a strong argument in favour of the Multi-Bang Hypothesis as opposed to the Big Bang Model. Scientific support for the Multi-Bang mode of galaxy formation is found in the following observations:

- The rotational curve of spiral galaxies shows stars and clouds of hydrogen moving at high velocity in the outer parts of these galaxies. This is readily understood if, according to the Multi-Bang Hypothesis, the impulse of the motion came from outside the galaxies themselves.

- The spiral arms of galaxies are visual evidence of the filaments of neutral hydrogen being drawn into galaxies from centres of the initial formation of matter.

- Cooling flows of hot hydrogen gas flowing towards the centres of clusters of galaxies may also represent filaments of neutral hydrogen gas produced by an initial mechanism.

It has come to light that galactic mergers also called galactic cannibalism may take place. Irregular galaxies which appear to be young may be the result of the galactic merger of older galaxies. The fact that galaxies can collide may be a proof of the Multi-Bang Hypothesis in that if galaxies were formed at points of intersection, they could be moving in different directions relative to each other and thus cross each others' path. The Big Bang Model would imply that galaxies would be moving outwards from a single point of explosion and would be unlikely to cross each others' path.

Dark galaxies found in a much earlier universe, detected only because of their absorption and lensing of light from more distant quasars are galaxy-sized clouds of atomic hydrogen. I have identified these as proto-galaxies, dark because they have not yet formed stars in molecular clouds. I envisage dark galaxies as being part of an evolution of galaxies.

The Multi-Bang Hypothesis has no need for dark matter composed of exotic subatomic particles to explain observations concerning the dynamics of galaxies. If there is no dark matter, then there is no need for dark energy to explain the continuing expansion of the universe.

The Multi-Bang Hypothesis allows the universe to be taken at face value to a much higher degree than the Big Bang Model allows. The cosmological vision proposed in the Theory of Nanocreation and Entropic Evolution is that the vastness of what is observed of the universe is what is out there; it does not represent a mere 2-5 % of what exists. I hold the opinion that in theories of the universe less is more.

Bibliography

Cassé, Michel (2003) *Stellar Alchemy: The Celestial Origin of Atoms*
Cambridge University Press

Encyclopedia Britannica 2011 Standard Edition: Black hole; Cosmos;
Hubble, Edwin Powell

Gribbin, John & Martin Rees (1991) *Cosmic Coincidences: Dark
Matter, Mankind and Anthropic Cosmology* Black Swan

Hawking, Stephen W. (1988) *A Brief History of Time: From the Big
Bang to Black Holes* Bantam Books

Zeilik, Michael (2002) *Astronomy: The Evolving Universe* 9th edition
Cambridge University Press

CONCLUSION to PART II

Nanocreation, cosmic evolution and the elements

The origin of the universe according to the Theory of Nanocreation is to be found in the creation of matter. The universe is essentially composed of hydrogen. The hydrogen atom is the simplest type of atom consisting of one proton orbited by one electron. The creation of the universe involved the creation of hydrogen.

The Theory of Nanocreation presents the view that the properties of matter are purpose-built and arise from the interactions of subatomic particles. The subatomic particles of which atoms are composed allow atoms to form by recombination; energy to be released from atoms and be absorbed by atoms; neutrons to form and become the constituents of new elements; and atoms to interact and become linked through chemical bonds to form molecules.

The properties of the subatomic components of atoms were designed such that a material universe could come into existence, develop spatial dimensions through expansion and have duration through time.

The transformations that the universe has undergone since its beginning constitute what is known as cosmic evolution. Cosmic evolution comprises the formation of galaxies in clusters and stars within the molecular clouds of galaxies. Cosmic evolution also comprises the stages that stars go through before ending their lives in nova and supernova explosions and becoming stellar corpses and black holes; as well as the formation of planets which orbit stars in planetary systems.

According to the Theory of Nanocreation, the fundamental heavy elements have been created as catalysts in star formation and as the components of life. It is proposed that unicellular life exists in vast quantities in association with dust in interstellar dark clouds and these carbon-based cells represent the first forms of life.

The fundamental elements have specific properties that serve the different functions of structure and metabolism within cells. The proportions of these elements in the universe reflect the roles they play in cells or their roles as catalysts. On planets such as Earth, a large proportion of fundamental elements which may once have been components of life, are now found incorporated into rocks or as gases in the atmosphere. A large proportion of the 1 % heavy elements in the universe represent the debris of life that has ceased living.

There are rare elements in the universe that, according to the view presented, have evolved from the fundamental elements by processes such as neutron capture during events such as supernova explosion. These elements are radioactive and decay into other more stable elements.

Therefore, in the view of Nanocreation, cosmic evolution does not involve the evolution of the principle heavy elements within stars by nuclear fusion reactions; cosmic evolution does, however, involve the evolution of elements from fundamental, created elements.

The Theory of Nanocreation describes a universe – only one universe – that is free. The universe was created according to a purpose hidden in God, but it is not guided in its details. The natural processes of cosmic evolution follow a natural course: galaxies, stars and planets are formed by natural processes and they disappear (eventually) by natural processes.

The universe itself has undergone change, and the structures within it undergo modification over time. The universe will run down, 'wear out' and 'grow old'. The destiny of the universe is one of the principle subjects of theology and sometimes philosophy; the answer to the question of purpose and destiny is beyond the material world; it points to the mystery approached only by faith.

Summary of hypotheses in Part II

The **Fundamental Elements – Components of Life Hypothesis** (Chapter 12) states that:

Fundamental elements (with the exception of hydrogen, oxygen, silicon and iron) were created as the constituents of living cells inhabiting interstellar dust clouds as unicellular forms of life. The properties of the fundamental elements were designed to sustain life. The fundamental elements found at the top of the Periodic Table of Elements have evolved into the elements found lower down in the chart, as well as into unstable isotopes of each element and inert gases by natural processes.

The elements of which all forms of life are composed number in total about 24:

Carbon (C), hydrogen (H), oxygen (O), nitrogen (N), phosphorus (P) and sulphur (S) make up the basic macromolecule cell constituents consisting of nucleic acids, proteins, carbohydrates and lipids.

Silicon (Si), calcium (Ca), phosphorus and fluorine (F) are employed in constructing the matrix of shells and skeletons.

Sodium (Na), potassium (K), calcium and chlorine (Cl) are involved in the physiology of cell function such as signalling. Chlorine is part of hydrochloric acid for digestion in animals.

The metal magnesium (Mg) and the transition metals manganese (Mn), iron (Fe), cobalt (Co), nickel (Ni), copper (Cu) and zinc (Zn) have properties crucial to the functioning of enzymes in plants or animals or both. Iron and copper are employed as oxygen-carriers in blood. Zinc is a component of DNA and acts in signalling in various ways including as a neurotransmitter in the brain of animals.

Selenium (Se) and iodine (I) are components of the thyroid hormones in animals.

The transition metals vanadium (V), tungsten (W) and molybdenum (Mo) are employed in the enzymes of anaerobic and aerobic bacteria, but not higher organisms.

I propose with the Fundamental Elements – Components of Life Hypothesis that the creation of unicellular life introduced 12 of these elements to the universe. These 12 elements were created from hydrogen atoms and their properties serve the functions performed by cells.

The bulk elements making up living cells are carbon, hydrogen, oxygen, nitrogen, phosphorus and sulphur; also, sodium, chlorine, potassium, calcium and magnesium. Apart from these 11, the other elements of life – mostly transition metals – are trace elements. These 11 bulk elements of life are found at the top of the Periodic Table of Elements and their abundance in the universe reflects their abundance as the components of cells (with the exception of hydrogen and oxygen).

In summary the elements of life include:

- Hydrogen – the basic stuff of the universe.
- Elements first created as catalysts (to be discussed shortly): oxygen, iron and silicon.
- Twelve elements created as the components of life: carbon, nitrogen, sodium, magnesium, phosphorus, sulphur, chlorine, potassium, calcium, manganese, copper and zinc.
- Various evolved elements including cobalt and nickel among others.

The division of the Periodic Table of Elements into fundamental elements and evolved elements accounts for the transformation of elements into new elements and into different isotopes of the same element by natural processes.

These natural processes (described in Chapter 12) include:

- Nuclear fusion of hydrogen inside the thermonuclear furnaces of stars produces the inert gas helium (He). The *PP chain* (proton-proton chain reaction) converts matter into energy; the resulting radiation causes stars to shine.

- Supernova synthesis involves rapid neutron capture followed by the transformation of neutrons into protons by beta decay

218

(β decay). This is the *r process*. Nucleosynthesis during the supernova explosions of red giant stars produces silver (Ag), gold (Au), platinum (Pt) and uranium (U).

- The unstable radioactive elements with atomic numbers greater than 82 decay into lead (Pb). An example of this is uranium. This involves a nuclear fission process and the emission of gamma rays.

- The light elements lithium (Li), beryllium (Be) and boron (B) at the top of the chart are produced by the nuclear fission of carbon and oxygen nuclei. Nuclear fission occurs when the cosmic ray nuclei of these elements collide at high speed with particles in the upper atmosphere or in interstellar dust clouds. The process is known as spallation.

- The stable elements have on average three isotopes which are atoms of the element with different numbers of neutrons but the same number of protons in the nucleus. Different isotopes of an element have different atomic masses, but the same atomic number. An element can be transformed into an isotope of a neighbouring element by neutron bombardment. ^{14}N nitrogen is transformed into the radioactive isotope of carbon ^{14}C by bombardment of the Earth's atmosphere by cosmic rays. ^{14}C is incorporated into plants along with ^{12}C. Carbon-14 decays back into nitrogen-14 with a half-life of 5717 years and this allows the radiometric dating of artefacts.

Hypothetical nuclear fusion reactions said to produce carbon, oxygen and silicon as well as other elements at the top of the Periodic Table by natural processes inside the cores of dying stars are not included under this scheme. The existence is not accepted of these processes said to involve extremely high temperatures of hundreds of millions of degrees inside stars with surface temperatures of the stars of only between 2000 and 6000 K.

The evolved elements include those that have evolved by the *r process* of rapid neutron capture: Iron (Fe 26) has evolved into cobalt (Co 27) and nickel (Ni 28); Magnesium (Mg 12) may evolve into aluminium (Al 13). Other evolved elements have arisen through radioactive decay: Potassium (K 19) decays into the inert gas argon (Ar 18).

219

With the exception of the inert gases, the evolved elements are mostly not very abundant in the universe. They generally do not have biological roles. Evolved elements are often unstable and radioactive, decaying into other elements.

Therefore, I believe that the fundamental elements found at the top of the Periodic Table have been created with specific properties to serve biological roles as the components of life. The fundamental elements have been modified by natural processes evolving into other elements which fill up the lower rows of the Periodic Table. Evolved elements show much less stability in their existence; the man-made elements often have a half-life of only split seconds.

The discovery of a resonance level in carbon-12 was taken as confirmation that carbon is produced inside dying stars by the fusion of helium in the triple alpha process. An energy level found in oxygen prevents carbon from transmuting into oxygen. Energy levels in elements may serve more in the prevention of transmutation of one element into another than in their production by nuclear fusion.

I contest that if God created carbon and other fundamental elements, they would *all* be created with energy levels such that they would not easily transmute into different elements. If the elements of life transformed themselves into one single element – iron for example – life would be wiped out and creation would be undone. The specific energy levels of fundamental elements prevent their transformation, except in violent astronomical events such as supernova explosion or cosmic ray bombardment. Thus, the energy level found in carbon is not proof that carbon was produced inside stars by nuclear fusion; if carbon were created as such, it would still have to have specific energy levels in order to remain as the element carbon.

The **Iron and Silicate Catalysts Hypothesis** (Chapter 13) states that:

The elements iron, silicon and oxygen were created as iron and silicate catalysts in the form of fine interstellar dust within the atomic hydrogen gas of galaxies. The catalytic properties of these elements allowed atomic hydrogen to be converted into molecular hydrogen forming molecular clouds. Stars form within molecular clouds by natural processes.

Stars are born in molecular clouds of H_2; thus, the question of how atomic hydrogen is converted into molecular hydrogen is central to understanding the processes of star formation. It is thought that interstellar dust grains play a catalytic role in molecule formation.

Interstellar dust grains tend to be about one micron (1 μm) in diameter, although many dust grains measure only a fraction of a micron, while some measure much more than a micron.

With the Iron and Silicate Catalysts Hypothesis, I propose that the smaller sized interstellar dust grains composed of silicates and iron were created in early galaxies as catalysts for the formation of molecular hydrogen. Thus, iron, silicon and oxygen are numbered among the fundamental elements and show high abundances – oxygen is ranked third and iron is ranked joint fourth with carbon in the abundance of elements in the universe.

I suggest that molecule formation rather than gravitational collapse initiates star formation. The formation of bonds when atoms become molecules may raise the density of matter, while the chemical energy of bond formation converted into kinetic energy may fuel the rotation of a molecular cloud. The rotation of a star comes from the rotation of the molecular cloud out of which it was born.

Chapter 13 contained a description of the evolution that stars undergo. Stars start off as protostars enshrouded in dust clouds; they mature into main sequence stars whose thermonuclear reactions cause them to shine; and they end their lives as stellar corpses, the larger ones exploding as supernovas sending shock waves through the galaxy.

In the current theory the elements are cooked up inside red giant stars, then blown out by stellar winds or supernova explosions to form nebulas out of which solar systems form. The elements carbon and silicon are often found associated with red giant stars.

I propose that carbon and silicon are found in the surface layers of red giant stars due to *lack* of a stellar wind in these dying stars when thermonuclear reactions in their interiors have ceased. The lack of stellar wind means that carbon and silicon flow into, or are engulfed by the red giant star causing it to become opaque and enter a phase of massive expansion. The carbon and silicates sampled in the surface layers of red giant stars are there because they have been drawn in from circumstellar dust rings, not because they have been made in the interior of the red giant and are being blown out.

In Chapter 11 I proposed that God created unicellular life forms in vast quantities within galaxies of the early universe. While it was iron and silicate dust which catalysed the formation of molecular hydrogen forming molecular clouds, it was the enzymes of unicellular life which catalysed the formation of complex carbon molecules and water found within dark molecular clouds.

This unicellular life may have included Archaea such as methanogens which are chemolithotrophs producing methane; Green and Purple bacteria which practice anoxygenic photosynthesis; Cyanobacteria which practice aerobic photosynthesis; and heterotrophic bacteria which live on organic matter. Alternatively, it may have included exotic types of Archaea and Eubacteria. Interstellar bacteria would be likely to have a filamentous form.

Therefore, the link between the creation of fundamental elements and the creation of life is to be found in interstellar forms of bacterial life.

Dark nebulas, the dust clouds of molecular clouds, show extinction of light from stars. I proposed in Chapter 11 that interstellar forms of life use silicate dust as miniature solar panels to harvest light for energy and anoxygenic photosynthesis. If this were so, interstellar dust grains may have an open structure like samples of dust from comets.

Comets appear to be giant snowballs that may have formed within interstellar dust clouds and be composed of interstellar dust grains with their icy mantles. Comets contain organic molecules and possibly viable life. (In Chapter 17 of Part III the capture of comets originating beyond the solar system is described as part of the Planet Capture Hypothesis). Likewise, meteorites may also have been captured. Carbonaceous chondrite meteorites appear to contain evidences of organic life, although this organic matter has been subjected to heat during transit through the Earth's atmosphere.

I have suggested that the gas giants of the solar system – Jupiter, Saturn, Uranus and Neptune – may be brown dwarfs which are failed stars grown old. These failed stars may have been captured into the solar system complete with their planetary rings of dust and ice. If this were so, planetary rings may represent a close-up view of interstellar dust and provide evidence of formerly thriving anaerobic interstellar life producing methane and water.

Although I believe that the carbon molecules found in interstellar dust clouds have been produced by life, many of the simpler molecules may represent break-down products from the decomposition of once-living cells. Elements belonging to the debris of life would undergo a constant recycling both in living metabolisms and physical processes.

The **Multi-Bang Hypothesis** (Chapter 15) states that:

Matter was created in the form of protons tightly packed together at saturation property density. Thus, the matter of the entire universe occupied a small volume. Created electrons were injected into the mass of protons at a myriad of points, possibly over a certain length of time. The high-energy electrons blew apart the proton mass in multiple bangs. Around each point, atoms of hydrogen formed. The natural recombination of a proton with an electron led to the formation of each atom. The space inside each atom became the cause of the cosmic expansion of the universe. The newly-formed atomic hydrogen gas streamed outwards from the points of expansion and met with other streams of hydrogen at intersections between the centres of expansion. The filaments of hydrogen curled round each other at intersections forming eddies; these eddies became galaxies formed from spiral arms of atomic matter.

In this vision of the beginning, cold, dark matter consisting of protons amassed together at the density of the atomic nucleus is blown apart by beams of electrons travelling close to or at the speed of light. Light floods into the darkness of the tiny dense universe.

At each point of input of energy, hydrogen atoms are formed from subatomic particles – protons and electrons. The space inside the atom brings about spacing between subatomic particles – this explosive expansion of atoms sets off the universal expansion. Time begins and space opens out. Moving matter creates spacetime – and this is the universe.

With the Multi-bang, expansion occurs from a myriad of points in all directions and atoms of hydrogen stream outwards. Where filaments of hydrogen met at the intersections between points of expansion, great eddies formed. Galaxies became the places where matter accumulated. The cords of matter became the spiral arms of galaxies.

223

Thus, galaxies formed through the interaction of matter in motion. The axis of rotation of a galaxy was determined by the motion of the dominant filament or filaments of matter of which a galaxy is composed gaining over that of the smaller filaments.

It had been widely assumed that the universe was infinite in extent and had always been there. The implication of expansion was that the universe had a beginning at a specific moment in the past. The new cosmology produced by scientific investigation revealed a universe that is vast, but finite in extent, and old but finite in duration. The universe had a beginning estimated to have occurred between 10 and 20 thousand million years ago; the figure most often cited for the age of the universe is 13.7 thousand million years.

The size of the universe is the age of the universe multiplied by the speed of light. Spacetime has a peculiar feature: looking out into space is looking back in time. Far-away objects existed in the early universe and do not exist now, yet the light they emitted thousands of millions of years ago is only just reaching us now. Astronomers with their powerful telescopes are blessed with the ability to look into and see the past happening now!

An explosive beginning to the universe followed by continuing expansion and cooling down is consistent with the Hot Big Bang Model and with the Multi-Bang Hypothesis.

In Chapter 15 I suggested that the centres of expansion could be identified with quasars. The tremendous energy released by quasars – a quasar can be brighter than a giant galaxy composed of a million million stars – makes them candidates for the Multi-Bang beginning as power-engines of the universe.

The cosmic background radiation (CBR) of the universe set out as ultraviolet light when the universe had a temperature of between 3000 and 4000 K. The photons are now in the microwave range and have a temperature of 2.735 K. According to the Multi-Bang Hypothesis the ultraviolet light that now forms the cosmic background radiation was sent travelling in all directions from the multiple points where atoms formed at the beginning of the universe.

Another cosmic background may be that of intergalactic primary cosmic rays. Primary cosmic rays travel at 87 % the speed of light which represents a velocity of 261 000 km/second and they come from all directions. I propose that the only credible mechanism for accelerating primary cosmic rays to this speed would be quasars with their synchrotron radiation travelling close to the speed of light. If this were so, primary cosmic rays would represent a fossil from the beginning of the universe and a confirmation of the Multi-Bang Hypothesis. The protons of primary cosmic rays would have come from the original proton mass and been accelerated by the explosive events of the Multi-Bang, while other protons recombined with electrons to form the first atoms.

Primary cosmic rays would have been travelling through intergalactic space for thousands of millions of years before entering a galaxy; they would not be the product of supernova explosions within the galaxy since supernovas can only eject material at a velocity of about 10 000 km/second which is far short of 261 000 km/second.

In Chapter 16 there was a description of galaxies and the stars they contain. It is estimated that the universe today contains 500 000 million galaxies. According to the current theory galaxies were formed by gravitational collapse in giant clouds of atomic hydrogen. According to the Multi-Bang Hypothesis galaxies were formed by matter (atomic hydrogen) in motion, and the motion came from multiple explosions external to the galaxies themselves. Galaxies formed at the intersections between centres of expansion as filaments of hydrogen became the spiral arms of galaxies.

The dominant opinion at the moment is that all galaxies are of the same age, all having formed some 10 000 million years ago. However, the speed of rotation of galaxies, the amount of neutral hydrogen gas they contain, and the age of stars within galaxies suggests that they can be classified into different age groups. The elliptical galaxies could be classified as old and the irregular galaxies as young, with the spiral galaxies in between. Galaxies classified by age appear to show that the spiral arms of galaxies are drawn into the central region which increases in size, while the spiral arms themselves become fainter, almost disappearing in old galaxies.

Although galaxies may be of different ages, they are no longer being formed now. The creation of matter in finite quantity at the beginning of the universe, and the expansion of the universe in size, would mean that the formation of new galaxies would cease completely at a certain point in time.

It is now known that a galaxy can swallow another galaxy. Irregular galaxies are either young galaxies or newly-produced galaxies from the galactic merger of older galaxies.

Galactic mergers may constitute a proof of the Multi-Bang Hypothesis in that, the fact that galaxies can collide shows that they are moving from points of intersection in all directions; they are not moving outwards from a single point. The Big Bang would have occurred at one point, whereas the Multi-Bang would have involved many points.

The dynamics of galaxies leads on to the subject of dark matter.

The rotational curves of galaxies show that stars in the outer parts of galaxies have rotational speeds as high as stars near the centres. This means that galaxies could not have formed by gravitational collapse or be maintained by gravity unless each galaxy had a halo of dark matter providing gravitational force that would boost the rotation of the outer parts of the galaxy.

The other reason for belief in the existence of undetected dark matter is that the Inflationary Big Bang Model requires a critical value for the total mass of the universe for rapid inflation to have slowed down. Since the observed matter in the universe only represents 10-20 % of the critical mass, it is hypothesized that dark matter makes up the missing mass in the universe.

Dark matter is thought to be composed of non-baryonic exotic particles. Particle physicists have set up experiments underground to catch WIMPS which stands for weakly interacting massive particles, but none have been detected so far.

In fact, the dynamic observed in galaxies strongly supports the Multi-Bang Hypothesis. In the Multi-Bang mode of galaxy formation, galaxies are formed by matter in motion and the impulse of the motion came from explosion rather than gravitational collapse. Since the momentum was generated outside the galaxies themselves, matter moves at the same speed in the outer parts of a galaxy as it does in the inner parts.

The Multi-Bang Hypothesis does not need the concept of dark matter to explain the dynamic of galaxies. Also, the universe of the Multi-

Bang does not undergo a phase of sudden inflation; therefore no extra matter is required for the stopping of inflation. The Multi-Bang Hypothesis takes the vast observable universe of atomic matter more or less at face value – hypothetical forms of dark matter and dark energy are not required.

Chapter 14 is about the structure of matter.

Two hundred subatomic particles are known although there are only 12 main subatomic particles; however, most of these are unstable. In fact, only five subatomic particles are stable and form the constituents of matter. They are Up Quarks and Down Quarks (forming protons and neutrons), electrons, neutrinos and antineutrinos.

It is of no real surprise that the universe is formed from matter with no detectable admixture of antimatter, since if this were not so, the universe would have ceased to exist in a puff of energy a very long time ago. The observed universe is asymmetrical in that it consists of only matter. The functioning of the universe is dependent upon asymmetry.

In the Theory of Nanocreation, matter was created by God out of nothing; there was never a phase when matter disappeared into nothing through annihilation with antimatter. God created matter with specific properties such that matter can interact and produce a stable, but evolving universe with duration through time.

Recap of hypotheses in Parts I and II

These nine hypotheses belong to the Theory of Nanocreation and Entropic Evolution:

The **Coding DNA Hypothesis** (Chapter 1) states that:
New genes at new gene loci were created as codes that carry information as the genetic program for life. New genes would be added to the genomes of the male and female gametes after fertilization and before DNA replication and syngamy. The new DNA codes would be added to the genome in its highest evolved state for each biological line without the deletion of previous genes or DNA.

The **Gene Tagging Hypothesis** (Chapter 2) states that:
Genes have been created with enhancers containing binding sites called regulatory elements for the binding of transcription factors that control gene expression. The Genetic Tagging System allows the execution of a genetic program. In plants and animals gene tagging allows the development of a multicellular body and maintains the body by coding for on-going metabolic processes.

The **Insertion of Middle Repetitive DNA Hypothesis** (Chapter 3) states that:
In the creation of new forms of life, enhancers with new regulatory elements were inserted into the genome packaged in middle repetitive DNA elements. Enhancers control the expression of new genes and existing genes and thereby produce new levels of morphological complexity. Repetitive DNA serving architectural roles also exists in the genome by design.

The **Ecological Constraint Hypothesis** (Chapter 8) states that: Ecosystems on Earth have been established by the introduction of new kinds of organisms with increasing levels of complexity through acts of creation over time. A new kind may only be introduced into an ecosystem, when the ecosystem provides it with an environment in which it can practice its new mode of life. For this reason, acts of creation are subject to ecological constraint in their timing and structural possibilities. Ecosystems can only be built-up layer by layer, by adding new classes of organisms with higher levels of complexity, at intervals that allow the adaptive radiation of each new class of organism to occur.

The **Genetic Switching-off Hypothesis** (Chapter 2) states that: Evolution occurs in the first step by random mutation causing genes or their regulatory elements to be switched-off. Genetic switching-off affects which genes are expressed and the timing of expression. When the switched-off version is advantageous in the phenotype, it is preserved by Natural Selection.

The **Life Produced Water Hypothesis** (Chapter 10) states that: Water, including the water found on Earth has originally been produced by the metabolic processes of life; most notably by anaerobic and aerobic chemolithotrophy, anoxygenic photosynthesis and anaerobic respiration in bacteria. Water is also produced by aerobic respiration in higher forms of life. Biosynthesis increases the amount of water in the environment where there is a rise in biomass.

The **Fundamental Elements – Components of Life Hypothesis** (Chapter 12) states that: Fundamental elements (with the exception of hydrogen, oxygen, silicon and iron) were created as the constituents of living cells inhabiting interstellar dust clouds as unicellular forms of life. The properties of the fundamental elements were designed to sustain life. The fundamental elements found at the top of the Periodic Table of Elements have evolved into the elements found lower down in the chart, as well as into unstable isotopes of each element and inert gases by natural processes.

The **Iron and Silicate Catalysts Hypothesis** (Chapter 13) states that:
The elements iron, silicon and oxygen were created as iron and silicate catalysts in the form of fine interstellar dust within the atomic hydrogen gas of galaxies. The catalytic properties of these elements allowed atomic hydrogen to be converted into molecular hydrogen forming molecular clouds. Stars form within molecular clouds by natural processes.

The **Multi-Bang Hypothesis** (Chapter 15) states that:
Matter was created in the form of protons tightly packed together at saturation property density. Thus, the matter of the entire universe occupied a small volume. Created electrons were injected into the mass of protons at a myriad of points, possibly over a certain length of time. The high-energy electrons blew apart the proton mass in multiple bangs. Around each point, atoms of hydrogen formed. The natural recombination of a proton with an electron led to the formation of each atom. The space inside each atom became the cause of the cosmic expansion of the universe. The newly-formed atomic hydrogen gas streamed outwards from the points of expansion and met with other streams of hydrogen at intersections between the centres of expansion. The filaments of hydrogen curled round each other at intersections forming eddies; these eddies became galaxies formed from spiral arms of atomic matter.

Entropy and a universe running down

Part I of *The Steps of Creation* contained an explanation of the Theory of Entropic Evolution as applied to life forms and biological evolution. In this context 'entropy' involved loss of information and the concept was applied to the transmission of information in the DNA code of life. The cosmic evolution described in Part II of *The Steps of Creation* cannot rightly be qualified as entropic; however, the concept of entropy has been widely applied to the universe as a whole. In this case, entropy concerns the dissipation of energy in a general running down of the universe. It is to this aspect that we will now turn.

The unquestioning view of the universe was that it has always been there, eternally as it is, infinite in extent. It was the formulation of the

Laws of Thermodynamics which showed this not to be the case. As mentioned in the Conclusion to Part I, the First and Second Laws of Thermodynamics were formulated during the 1840s and 1850s by a number of physicists. The contributions of these physicists are described in an appendix to this Conclusion.

The classic applications of the Laws of Thermodynamics concern energy within closed systems. The universe as a whole may be taken as a closed system.

Applied to the universe, the First Law of Thermodynamics states that the energy of the universe is constant. This means that energy can neither appear nor disappear in the universe; energy can only be converted into another form of energy. All the energy contained in the universe must have been supplied at the beginning and from then on remained constant.

The Second Law of Thermodynamics states that it is impossible to convey heat from a body of lower temperature to another at a higher temperature. It was observed that when two bodies at different temperatures were placed in contact, there is a transfer of heat from the warmer to the cooler body without work being done. This means that energy is dissipated as heat, and this heat cannot be used for mechanical work. It was found that all natural processes have a tendency towards the dissipation of energy.

The concept of entropy is part of the Second Law of Thermodynamics. An increase in entropy is an increase in disorder, and this only occurs in one direction; things only naturally become more disorderly, they do not build up order without an external input of energy and information.

The Second Law of Thermodynamics states that within a system there is an inevitable degradation of energy and an increase in entropy as a consequence of random events. Even if entropy decreases (and order is built-up) in parts of a system, there will still be a corresponding overall increase in entropy of the whole system.

Disorder is the degree of randomness of a system – randomness produces a statistically even distribution of matter and temperature in a system. A disordered system is observed to be highly homogenous.

When entropy has reached its final conclusion, a disordered system is one in which no difference of potential remains, and in which no macroscopic phenomena can occur. The system has become inert. In this sense the Second Law of Thermodynamics proclaims that systems run down, and in its application to the universe, it means that the universe is running down.

Therefore, according to the Second Law of Thermodynamics, an isolated system will eventually reach internal thermal equilibrium, after which time only random fluctuations about the state of equilibrium can take place. The system can never return to its former non-equilibrium state.

One conclusion that may be drawn from the Laws of Thermodynamics is that the world is subject to decay. The universe is running down; it will advance towards 'old age'.

The second conclusion to be drawn from the Laws of Thermodynamics is that there is an arrow of time – time moves in one direction only. Time moves from a beginning to an end because of the irreversibility of natural processes. The dissipation of energy means that there is no perpetual motion and the universe is not caught in an eternal cycle of processes.

The third conclusion presented as the Entropic Creation Argument is that the universe had a beginning.

Entropic Creation
The universe viewed as a system is very far from equilibrium; it contains order, structure, life and available energy. If the entropy of the universe is increasing as free energy is running down, it cannot have been running down forever. The universe must have come into existence sometime in the past with sufficient available energy for entropy not to have reached its maximum value yet.

The implication of the Second Law of Thermodynamics is that the present non-equilibrium universe cannot be of infinite age. It must have had a beginning when it had minimum entropy or maximum free energy. This is the essence of the entropic argument for creation.

Helge Kragh (2008)[25], professor of the history of science at the University of Aarhus, Denmark encapsulates the Entropic Creation Argument as follows:

".... given that we do not live in a high-entropic world, and assuming that the entropy law is valid for the universe at large, entropy can only have increased for a finite period of time. If this is the case, the universe must have had a beginning of a sort – and if it had a beginning it presumably originated in a creative act. This is the essence of the 'entropic creation argument' " (Kragh 2008, page 1).

This scientific indication that the universe is finite and had a beginning was used by Theists in the 19th century to counter the belief in Materialism put forward by Atheists at that time. It also became the setting for the 20th century advances in cosmology concerning the origin of the universe at a point in time.

Heat Death
When all the energy of the universe is transformed into heat, and all the heat of the universe reaches a state of equilibrium, then all possibility of further change will have come to an end, and the complete cessation of natural processes must occur.

The conclusion is that when the entropy of the universe tends towards its maximum, the universe will be condemned to what Helmholtz described as a state of eternal rest, and what Clausius described as a state of absolute rest or universal death.

The predicted future end to the universe is known as the Heat Death.

The Heat Death scenario does not proclaim an end to the universe itself; it does not involve the destruction of matter or energy, but a distribution of energy such that no further change is possible in the universe. The universe becomes one equally heated, inert mass. In the Heat Death, the universe runs down to a state where there is no free energy to sustain motion or life.

The Heat Death final end was applied to a universe thought only to consist of the stars of the Milky Way. There were many attempts to

[25] Kragh, Helge S (2008) *Entropic Creation: Religious Contexts of Thermodynamics and Cosmology* Ashgate Publishing Ltd

get round the gloomy inevitable end. Traditionally Christianity has held the view that God will bring this world to its conclusion before it wears out completely. According to this belief, reality will be transfigured in some way that nobody yet understands. In this way, the timeline of the universe will be cut short before the Heat Death occurs.

Einstein and the Laws of Thermodynamics

During the 19th century the highly improbable state of high free energy and low entropy at the beginning of the universe was explained by the notion that the universe started off as dispersed gaseous matter with energy being injected into this primordial matter to cause it to start to form stars. Theists such as William Thomson and James Clerk Maxwell who propounded the Laws of Thermodynamics believed that their science indicated that the energy of the universe had been supplied in the beginning by the Creator. The First Law of Thermodynamics showed that the universe could not have obtained energy from itself.

During the 20th century, Einstein's Theory of Special Relativity modified the First Law of Thermodynamics as applied to the universe in that energy became mass-energy. The equivalence between matter and energy was established with the famous formula $E = MC^2$. It was no longer necessary to create a universe by injecting energy into matter, since matter and energy were no longer viewed as separate things.

Quantum Theory shows that energy comes out of matter, and the Big Bang Standard Model proposes that the origin of matter was as energy.

The General Theory of Relativity showed that the extension of space only has meaning in connection with matter; there is no space beyond matter. Spacetime is measured by light emitted from matter and absorbed by matter; space exists in between matter, and it's spatial and time lapse measurement is interchangeable. This suggests a finite universe.

Just as the Laws of Thermodynamics cannot be applied to a universe infinite in time or extent, gravitation cannot be calculated in an infinite universe.

The model of the universe that emerged during the 20th century was very different from the one envisaged during the 19th century. The Laws of Thermodynamics still apply to the new model of the universe, but not quite in the same way.

Forms of energy and the Multi-Bang

Entropy is increasing in the universe as potential energy is converted into kinetic energy, and this energy is converted into heat. The dissipation of energy in the universe is the cause of a cosmic winding-down.

According to the Multi-Bang Hypothesis, God created matter. It was not necessary to create matter and put energy into it because energy can be obtained from matter.

The creation of protons followed by the creation of electrons at a myriad of points created potential energy. Potential energy is the energy of position, and it arises from the structure and distribution of matter. The initial quantity of matter created in the universe was important since it affects the working of gravity. The potential energy created from spatial position had to be sufficient to set the universe in motion.

As atoms formed, space was created. Streams of atomic hydrogen moving outwards formed galaxies, and transformed potential energy into kinetic energy. Kinetic energy is the energy of motion.

Molecules formed within swirling galaxies, and molecular clouds gave rise to stars. Within galaxies various processes combined with gravity transformed kinetic energy into thermonuclear energy – the heat and light of stars.

Energy in the universe is being converted into dispersed heat and the universe is cooling down due to cosmic expansion.

The space created inside the atom gave rise to an early universe characterized by fast-moving matter. This early matter is now represented by primary cosmic rays travelling at 87 % the speed of light. The initial fast-moving matter of the early universe has been superseded by matter moving at a more sedately pace. Even so the

235

universe still displays a high degree of order and is very far from reaching maximum entropy.

Standard Models and the Laws of Thermodynamics

The Big Bang is a theory which arose in the 20th century as a product of the Theory of General Relativity and observations concerning the recession of galaxies which showed that the universe is expanding. The origin of elements was at one time attributed to the inferno of the Big Bang explosion, although now only the production of helium is attributed to this stage.

A new idea entered the debate in the 1970s; it was the idea that the universe had not only begun in a large explosion, but also from a Big Bang Singularity – this concept proclaimed that the entire universe started off as a mathematical point.

The presently accepted Standard Model of the Big Bang presents the universe as having started out as a fireball of energy at a temperature of 10^{12} K (or one million million degrees) near the time 10^{-43} seconds. In this model the Big Bang explosion consisted of photons which as they expanded outwards, dropped in temperature causing energy to condense into matter and antimatter.

The Standard Model of the Big Bang became modified during the 1980s into the Inflationary Big Bang Model. The proposal that the universe evolved from a region much smaller (10^{50} times smaller) than that proposed in the Standard Big Bang Model, resolved the problem of obtaining a universe with an even distribution of temperature. The size was calculated from the speed of light. In the Inflationary Big Bang Model there is sudden inflation between 10^{-43} and 10^{-32} seconds bringing the universe from smaller than a proton to the size of a grapefruit; then the Standard Model of the Big Bang resumes. The temperature near the beginning of the Inflationary Big Bang Model is proposed to have been as much as 10^{32} K (or one hundred million million million million million degrees).

The Standard Model of Particle Physics describes the interaction between subatomic particles in terms of the exchange of massless intermediary particles. The chief of the hypothetical particles is the

Higgs boson postulated as carrying the Higgs field. It is hypothesized that through its interaction with other subatomic particles, the Higgs boson endows all elementary particles with mass. The mass of each of the known subatomic particles would occur, according to this theory, through the differing strengths of interaction of these particles with the Higgs field.

The link between the Standard Model of the Big Bang and the Standard Model of Particle Physics is that it is hypothesized that at the very high energy levels in the first instants of the Big Bang there was symmetry between matter and antimatter. Matter and antimatter annihilate each other instantaneously and nothing is left – no universe. The Higgs particle is invoked to break the hypothetical symmetry at the beginning, thus producing a universe of matter. The role of the Higgs boson in creation has led it to being dubbed 'the God particle'.

The Higgs particle is hypothesized as being a very massive particle, larger than a proton with a corresponding energy level that up to now could not be produced in particle accelerators. Advances in technology have only recently (2011) made it possible to realistically search for the Higgs particle; it is expected to be found or not found soon. Some believe that it has been found.

The hypothetical Higgs particle is possibly, for those who believe in it, pantheism – the belief that God is part of the material universe. The Creator God of Theism who is beyond, and greater than the material universe, does not need to create a universe that will rapidly annihilate itself, and then create a particle that will prevent it from doing so.

The view presented in the Multi-Bang Hypothesis is that there is no universe without matter. The first photons of light were released from matter. The proclamation in the Standard Model of the Big Bang that the universe started off as pure energy or a burst of light in the absence of matter is a contradiction according to this view.

The Standard Model of the Big Bang and Particle Physics proclaims that the very early universe was an almost uniform expanding ball of energy, and that matter was built up from this energy. The laws of entropy proclaim that the universe will end as homogenous energy. The existence of evenly distributed heat energy at the Big Bang prior to potential energy and kinetic energy – both of which cannot exist without matter, appears also to contravene the Laws of Thermodynamics.

Space has no meaning or existence in the absence of matter since it is the distance between bodies of matter measured by the speed of light that creates the three dimensions of space and that of time. There is no spacetime in the absence of matter.

God created matter

The Multi-Bang Hypothesis proclaims that God created matter. God created the universe *ex nihilo* – out of nothing. (According to this notion, there is a realm not dependent on the material realm which exists according to the will of God).

The known physical universe contains an estimated 10^{80} elementary particles (Dembski 2002, page 21)[26]. It is proposed in the Multi-Bang Hypothesis that all the particles of the universe were created in one locality at saturation property density. The protons created had a temperature at or close to absolute zero. The electrons created because of their high velocity, moving close to the speed of light had very high energy. Atoms formed at 3000-4000 K so this was the temperature of the universe in its initial stages.

The universe was formed by matter in motion. God set the amount of matter contained in the universe such that gravity would act within an on-going universe with extension and duration through time.

The universe was created by God according to a plan. A plan involves self-imposed constraint. In this way the universe created was governed by order and is open to rational inquiry.

God created matter at the nanoscale with specific intrinsic properties. All the macrostructures of the universe evolved from this microscopic matter, by various interactions involving natural processes.

---

[26] Dembski, William A. (2002) *No Free Lunch: Why Specified Complexity Cannot Be Purchased Without Intelligence* Rowman & Littlefield Publishers, Inc.

The currently accepted interpretation of Quantum Mechanics claims that at the subatomic level inside the atom there is a fundamental indeterminacy in the behaviour of subatomic particles (their location and velocity). Quantum Uncertainty has become a view of reality. Some have claimed that God can only act in the universe because matter is not completely determined by natural law. I believe that what for science is still uncertain is not uncertain for God. God does not require things to be uncertain or indeterminate at the quantum level to be able to act within the universe.

God acts in a quantum world that is known to Himself. An act of creation would occur through the reordering of atoms and molecules using energy – thus, the quantum world could be seen as providing the seam between nature and divine action.

The properties of matter were created such that interactions involving matter were possible. The properties of matter arise from the specific properties of subatomic particles. The interactions of matter in the universe may be described by the laws of physics and chemistry, but God did not create laws for matter, He simply created matter.

The chronological steps of creation

There are six steps of creation plus a seventh. Each step of creation involved a number of acts of creation. These are the chronological steps of creation:

1. Creation of protons. These subatomic particles of the nucleus are composed of up and down quarks. All the protons of the universe were created in one location at saturation property density.

     o *There was darkness before spacetime opened up.*

2. Creation of electrons. Electrons created at a myriad of points within the proton mass recombined to form the first atoms and produced the Multi-Bang. This blew apart the dense proton mass and created an expanding universe. Matter set in motion led to the formation of galaxies composed of atomic hydrogen, while primary cosmic rays started their journey through outer space.

     o *Light from points of expansion (quasars) flooded the early universe.*

3. Creation of heavy element catalysts as iron and silicate interstellar dust within galaxies. Molecular clouds of hydrogen formed within galaxies from which stars were born. Dying stars forged planets and planets were captured into planetary systems around other stars [27].

     o *Stars started to shine with light from thermonuclear reactions; galaxies containing thousands of millions of stars lit up the darkness of the expanding universe.*

4. Creation of living cells with the fundamental elements created as the components of life. Unicellular life was created within interstellar dust clouds in vast quantities; these life forms became associated with interstellar dust grains. These first cells were created with the DNA code of the prokaryotic genome.

---

[27] The explanation for this is found in Chapter 17 of Part III.

      o   *Water was produced in interstellar clouds and on the surfaces of planets (often found as ice). Bacteria started to transform molecules using enzymes as catalysts.*

5. Creation of protists on Earth. These single and multi-cellular eukaryotic organisms were created with the DNA code of the eukaryotic genome – a genome composed of chromosomes.

      o   *Protists transformed the Earth contributing oxygen to an aerobic atmosphere and forming sedimentary rocks that became the Earth's crust with its landmasses.*

6. Creation of multicellular plants and animals on Earth. Multicellular organisms have new modes of function in the genome [28]. Successive classes of plants and animals were created by the addition of coding DNA to the genomes of the various biological lines. New levels in the complexity of life were created as limited editions as ecological conditions allowed, culminating in the creation of humans.

      o   *There was a succession of ecological systems on Earth testified to by the fossils of the geological record.*

7. Creation of the human soul. Each human being conceived is given a soul. The soul is associated with the body and leaves the body when the person dies. The soul makes a spiritual life possible, and allows us to relate to God, our Creator.

      o   *The soul enters the next life in a way we do not fully understand, but hints are revealed to us in many religions.*

---

[28] The new modes of function in the eukaryotic genome of multicellular organisms involve middle repetitive DNA elements, alternative splicing of RNA and heterochromatin repression of gene expression.

Each of the steps of creation involved acts of creation. The number of acts of creation belonging to each successive step increases, while the amount of physical matter involved decreases.

The creation of protons involved one act of creation; the creation of electrons may also have involved the creation of other subatomic particles such as neutrinos and antineutrinos; the creation of interstellar dust involved three elements; the creation of interstellar life involved twelve new elements; the creation of successive classes of organisms on Earth involved about 120 acts of creation with about 1000 new genes for each new biological line; the creation of souls involves all the individuals who have lived on Earth which is thousands of millions.

The quantities of physical matter involved in each successive step of creation are less than in the previous step. The first step of creation involved almost all of the mass of the universe – protons and their conversion to neutrons which form the atomic nuclei of all the atoms in the universe. This was followed by the creation of electrons; the electron has a mass 1800 times less than that of the proton. The creation of heavy elements as interstellar dust involved only 1 % of the matter in the universe. The creation of unicellular life within interstellar clouds involved only the carbon content of these clouds together with a variety of other elements as the components of life. The creation of unicellular and multicellular life on Earth would only represent a tiny proportion compared to the unicellular life in interstellar clouds and on other planets throughout the universe. The molecules of life were formed from the already existing elements available. The soul is not composed of matter, although there appears to be a type of energy associated with the soul.

Each act of creation involving physical matter occurred at the nanoscale. The acts of creation involved firstly subatomic particles, then atoms and later molecules (such as the DNA of cells).

APPENDICES

Fundamental and linked elements

The fundamental elements appear to undergo modifications producing other elements in groups of three. The parent element gives rise to two other elements, which later decay back to the parent element.

Titanium-44 has a lifetime of 100 years. It decays to Scandium-44 then calcium-44 by emitting gamma rays.

Atomic number: **Ca** 20   Sc 21   Ti 22

Potassium (K) decays to calcium and the inert gas argon.

Atomic number: Ar 18   **K** 19   Ca 20

Magnesium and aluminium are linked. The radioactive isotope aluminium-26 decays to magnesium-26. This is a group of only two.

Atomic number: **Mg** 12   Al 13

Nitrogen can be converted into oxygen by bombardment. Nitrogen-14 also produces carbon-14 by cosmic ray bombardment in the atmosphere.

Atomic number: C 6   **N** 7   O 8
Atomic mass: 14.0067

Hydrogen can be transformed into two isotopes, deuterium is a stable isotope and tritium is unstable. Deuterium has one proton and one neutron. Tritium has one proton and two neutrons.

Atomic number: **H** 1   D   T or ^{3}H
Atomic mass: 1.00794

APPENDIX FOR CONCLUSION II

The First and Second Laws of Thermodynamics

Thermodynamic Theory was preceded by investigations into the functioning of steam engines. Most of the work on steam engines was advanced in Britain, however, it was the French scientist Sadi Carnot (1796-1832) whose paper published in 1824 became the basis to a theoretical understanding of heat engines. Carnot discovered that the efficiency of an idealized engine depends only on the temperature of its hottest and coldest parts and not on the substance used to drive the mechanism. These results are applied in engineering as the Carnot Cycle. Carnot's work was later incorporated into Thermodynamic Theory by Rudolf Clausius and William Thomson.

James Prescott Joule (1818-1889), an English physicist, established that the various forms of energy – mechanical, electrical and heat – are basically the same and can be changed one into another. This is the basis of the First Law of Thermodynamics – the conservation of energy.

The Laws of Thermodynamics were first formulated by the German mathematical physicist Rudolf Clausius (1822-1888). In 1850 Clausius presented a paper in which he presented two fundamental principles: the first principle was energy conservation or the equivalence of heat and work, and the second principle was that it is impossible to convey heat from a body of lower temperature to another at a higher temperature. Clausius applied the results of his experiments to the theory of the steam engine.

The Laws of Thermodynamics were also formulated by the Scottish physicist William Thomson, Lord Kelvin (1824-1907). In an essay published in 1851, Thomson expanded on James Joule's theory that heat is a form of motion, rejecting the accepted view of the time that heat was a substance called caloric.

Thomson explained that when two bodies at different temperatures are placed in contact, there is a transfer of heat from the warmer to the cooler body without work being done. The result of this is that although the energy is conserved, it has become dissipated in the sense that the system's capacity to perform mechanical or other work has diminished. The dissipation of energy involves the conversion of energy into heat.

The concept of entropy was added to the Second Law of Thermodynamics by Clausius in 1865. Max Planck reformulated the Second Law in terms of entropy increase in the late 1870s.

During the 1850s, the Scottish physicist James Clerk Maxwell (1831-1879) was working on the Mechanical Theory of Heat. His work was continued and given a mathematical formulation as the Dynamical Theory of Gases by Ludwig Boltzmann, an Austrian physicist, in the 1870s. Maxwell and Boltzmann both became convinced that entropy reflects molecular randomness. They showed that the spontaneous mixing of hot and cold gases is an irreversible process.

Maxwell and Boltzmann showed that a system with increasing entropy is a system in which the molecules of gases intermingle, differences in temperature disappear and differences in potential are evened out. Once statistical equilibrium is reached within the system, no further macroscopic phenomena will occur within it.

In 1882 Hermann von Helmholtz (1821-1894), a German scientist and philosopher, added the concept of free energy and bound energy. An increase in entropy corresponds to a decrease in free energy.

In summary, the First Law of Thermodynamics states that the energy of the universe is constant. This means that energy can neither appear nor disappear within a system; it can only be converted into another form of energy.

The Second Law of Thermodynamics states that it is impossible to convey heat from a body of lower temperature to another at a higher temperature. This means that nature has a tendency towards the dissipation of energy. Stated informally the Second Law means that in any process some energy is always dissipated as heat. When energy has been converted into heat, it cannot be used for mechanical work.

In physics, mechanical work is the measure of energy transfer that occurs when an object is moved over a distance by an external force. Work is obtained from order.

Bibliography

Encyclopedia Britannica 2011 Standard Edition: Carnot, Sadi; Clausius, Rudolf (Julius Emanuel); Entropy; Helmholtz, Hermann von; Joule, James Prescott; Kelvin, William Thomson, Baron; Maxwell, James Clerk; Work.

ABOUT THE AUTHOR

The author studied anthropology and sociology at Oxford Brookes University in England and Aix-en-Provence University in France obtaining a BSc Honours degree in 1984, a 'Licence' in 1988, and a 'Maîtrise' (masters degree) in 1990.

She is a practicing Christian attending the local Roman Catholic Church.